Lecture Notes in Mathematics

A collection of informal reports and seminars
Edited by A. Dold, Heidelberg and B. Eckmann, Zürich

43

Daniel G. Quillen

Massachusetts Institute of Technology
Cambridge, Mass.

Homotopical Algebra

1967

Springer-Verlag · Berlin · Heidelberg · New York

Library of Congress Catalog Card Number 67-29950. Title No. 7363

Homotopical Algebra

Daniel G. Quillen[1]

Homotopical algebra or non-linear homological algebra is the generalization of homological algebra to arbitrary categories which results by considering a simplicial object as being a generalization of a chain complex. The first step in the theory was presented in [5], [6], where the derived functors of a non-additive functor from an abelian category $\underline{A}$ with enough projectives to another category $\underline{B}$ were constructed. This construction generalizes to the case where $\underline{A}$ is a category closed under finite limits having sufficiently many projective objects, and these derived functors can be used to give a uniform definition of cohomology for universal algebras. In order to compute this cohomology for commutative rings, the author was led to consider the simplicial objects over $\underline{A}$ as forming the objects of a homotopy theory analogous to the homotopy theory of algebraic topology, then using the analogy as a source of intuition for simplicial objects. This was suggested by the theorem of Kan [10] that the homotopy theory of simplicial groups is equivalent to the homotopy theory of connected pointed spaces and by the derived category ([9], [19]) of an abelian category. The analogy turned out to be very fruitful, but there were a large number of arguments

---

[1]Supported in part by the National Science Foundation under grant GP 6166.

which were formally similar to well-known ones in algebraic topology, so it was decided to define the notion of a homotopy theory in sufficient generality to cover in a uniform way the different homotopy theories encountered. This is what is done in the present paper; applications are reserved for the future.

The following is a brief outline of the contents of this paper; for a more complete discussion see chapter introductions. Chapter I contains an axiomatic development of homotopy theory patterned on the derived category of an abelian category. In Chapter II we give various examples of homotopy theories that arise from these axioms, in particular we show that the category of simplicial objects in a category <u>A</u> satisfying suitable conditions gives rise to a homotopy theory. Also in §5 we give a uniform description of homology and cohomology in a homotopy theory as the "linearization" or "abelianization" of the non-linear homotopy situation, and we indicate how in the case of algebras this yields a reasonable cohomology theory.

The author extends his thanks to S. Lichtenbaum and M. Schlesinger who suggested the original problem on commutative ring cohomology, to Robin Hartshorne whose seminar [9] on Grothendieck's duality theory introduced the author to the derived category, and to Daniel Kan for many conversations during which the author learned about simplicial methods and formulated many of the ideas in this paper.

Contents

Chapter I. Axiomatic Homotopy Theory

Introduction

Chapter I is an attempt to define what is meant by a "homotopy theory" in a way sufficiently general for various applications. The basic definition is that of a model category which is a category endowed with three distinguished families of maps called cofibrations, fibrations, and weak equivalences satisfying certain axioms, the most important being the following two: M1. Given a commutative solid arrow diagram

$$\begin{array}{ccc} A & \longrightarrow & X \\ {\scriptstyle i}\downarrow & \nearrow & \downarrow{\scriptstyle p} \\ B & \longrightarrow & Y \end{array}$$

(the diagonal arrow $B \to X$ is dotted)

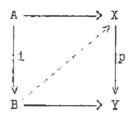

where i is a cofibration, p is a fibration, and either i or p is also a weak equivalence, there exists a dotted arrow such that the total diagram is commutative. M2. Any map f may be factored $f = pi$ and $f = p'i'$ where i, i' are cofibrations where p, p' are fibrations, and where p and i' are also weak equivalences. It should be noticed that we do not assume the existence of a path or cylinder functor; in fact the homotopy relation for maps may be recovered as follows: Call an object X cofibrant if the map $\emptyset \to X$ is a cofibration (hence in the category of simplicial groups the cofibrant objects are the free simplicial groups) and <u>fibrant</u> if the map $X \to e$ is a fibration (hence in the category of simplicial sets the fibrant objects are the Kan complexes). Then two maps f, g from a cofibrant object A to a fibrant object B are said to be <u>homotopic</u> if there exists a commutative diagram

(1)

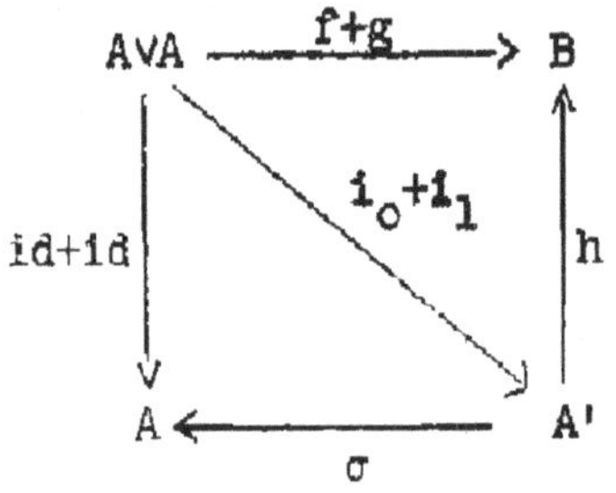

where $\vee$ denotes direct sum, $f+g$ is the map with components f and g, and where σ is a weak equivalence.

Given a model category $\underline{C}$, the <u>homotopy category</u> $\mathrm{Ho}\,\underline{C}$ is obtained from $\underline{C}$ by formally inverting all the weak equivalences. The resulting "localization" $\gamma : \underline{C} \to \mathrm{Ho}\,\underline{C}$ is in general not calculable by left or right fractions [7] but is rather a mixture of both. The main result of §1 is that $\mathrm{Ho}\,\underline{C}$ is equivalent to the category $\pi\underline{C}_{cf}$ whose objects are the cofibrant and fibrant objects of $\underline{C}$ and whose morphisms are homotopy classes of maps in $\underline{C}$. If $\underline{C}$ is a pointed category then, in §§2-3 we construct the loop and suspension functors and the families of fibration and cofibration sequences in the homotopy category. If one defines a <u>cylinder object</u> for a cofibrant object A to be an object A'

together with a cofibration i_0+i_1 and a weak equivalence σ as in diagram (1), then the constructions are the same as in ordinary homotopy theory except that, since a cylinder object of A is neither unique nor functorial in A, one has to be careful that things are well-defined. This is done by defining operations in two ways using the left (cofibration) structure and the right (fibration) structure, and showing that the two definitions coincide.

The term "model category" is short for "a category of models for a homotopy theory", where the homotopy theory associated to a model category $\underline{C}$ is defined to be the homotopy category Ho $\underline{C}$ with the extra structure defined in §§2-3 on this category when $\underline{C}$ is pointed. The same homotopy theory may have several different models, e.g. ordinary homotopy theory with basepoint is ([10], [15]) the homotopy theory of each of the following model categories: 0-connected pointed topological spaces, reduced simplicial sets, and simplicial groups. In section 4 we present an abstract form of this result which asserts that two model categories have the same homotopy theory provided there are a pair of adjoint functors between the categories satisfying certain conditions.

This definition of the homotopy theory associated to a model category is obviously unsatisfactory. In effect, the loop and suspension functors are a kind of primary structure on Ho $\underline{C}$, and the families of fibration and cofibration sequences are a kind of secondary structure since they determine the Toda bracket (see §3) and are equivalent to the Toda bracket when Ho $\underline{C}$ is additive. (This last remark is a result of Alex Heller.) Presumably there is higher order structure ([8], [17]) on the homotopy category which forms part of the homotopy theory of a model category, but we have not been able to find an inclusive general definition of this structure with the property that this structure is preserved when there are adjoint functors which establish an equivalence of homotopy theories.

In section 5 we define a closed model category which has the desirable property that a map is a weak equivalence if and only if it becomes an isomorphism in the homotopy category.

Chapter I. Axiomatic Homotopy Theory.

§1. The Axioms.

All diagrams are assumed to be commutative unless stated otherwise.

Definition 1: By a model category we mean a category together with three classes of maps in $\mathcal{C}$, called the fibrations, cofibrations, and weak equivalences, satisfying the following axioms.

M0. $\mathcal{C}$ is closed under finite projective and inductive limits.

M1. Given a solid arrow diagram

(1)

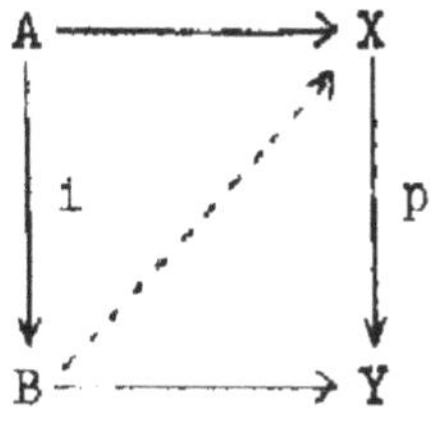

where i is a cofibration, p is a fibration, and where either i or p is a weak equivalence, then the dotted arrow exists.

M2. Any map f may be factored $f = pi$ where i is a cofibration and weak equivalence and p is a fibration. Also $f = pi$ where i is a cofibration and p is a fibration and weak equivalence.

M3. Fibrations are stable under composition, base change, and any isomorphism is a fibration.

Cofibrations are stable under composition, co-base change, and any isomorphism is a cofibration.

M4. The base extension of a map which is both a fibration and a weak equivalence is a weak equivalence. The co-base extension of a map which is both a cofibration and a weak equivalence is a weak equivalence.

M5. Let $X \xrightarrow{f} Y \xrightarrow{g} Z$ be maps in $\mathcal{C}$. Then if two of the maps f, g, and gf are weak equivalences, so is the third. Any isomorphism is a weak equivalence.

Examples. A. Let $\mathcal{C}$ be the category of topological spaces and continuous maps. Let fibrations in $\mathcal{C}$ be fibrations in the sense of Serre, let cofibrations be maps having the lifting property of Axiom M1 whenever p is both a Serre fibration and a weak homotopy equivalence, and finally let weak equivalences in $\mathcal{C}$ be weak homotopy equivalences (maps inducing isomorphisms for the functions $[K,\cdot]$ where K is a finite complex). Then the axioms are satisfied. (This is proved in Chapter II, §3.)

B. Let $\mathcal{Q}$ be an abelian category with sufficiently many projectives and let $\mathcal{C} = C_+(\mathcal{Q})$ be the category of complexes $K = \{K_q, d: K_q \to K_{q-1}\}$ of objects of $\mathcal{Q}$ which are bounded below ($K_q = 0$ if $q \ll 0$). Then $\mathcal{C}$ is a model category where weak equivalences are maps inducing isomorphisms on homology, where fibrations are the epimorphisms in $\mathcal{C}$, and where cofibrations are maps i which are injective and such that Coker i is a complex having a projective object of $\mathcal{Q}$ in each dimension.

C. Let $\mathcal{C}$ be the category of semi-simplicial sets and let fibrations in $\mathcal{C}$ be Kan fibrations, cofibrations be injective maps, and let the weak equivalences be maps which become homotopy equivalences when the geometric realization functor is applied. Then $\mathcal{C}$ is a model category (Ch II, §3).

For the rest of this section $\mathcal{C}$ will denote a fixed model category.

Definition 2: Let $\emptyset$ (resp. e) denote "the" initial (resp. final) object of the category $\mathcal{C}$. (These exist by M0.) An object X will be called cofibrant if $\emptyset \to X$ is a cofibration and fibrant if $X \to e$ is a fibration. A map which is both a fibration (resp. cofibration) and a weak equivalence will be called a trivial fibration (resp. trivial cofibration.)

Remark: In example A every object is fibrant and the class of cofibrant objects includes CW complexes, and more generally any space that constructed by a well ordered succession of attaching cells. In example B every object is fibrant and the cofibrant objects are the projective complexes (that is, complexes consisting of projective objects--these are not projective objects in $C_+(\mathcal{Q})$). In example C every object is cofibrant and the fibrant objects are those s.s. sets satisfying the extension condition.

Before stating the next definition we recall some standard notation concerning fibre products and introduce some not-so-standard notation for cofibre products. Given a diagram

(2)

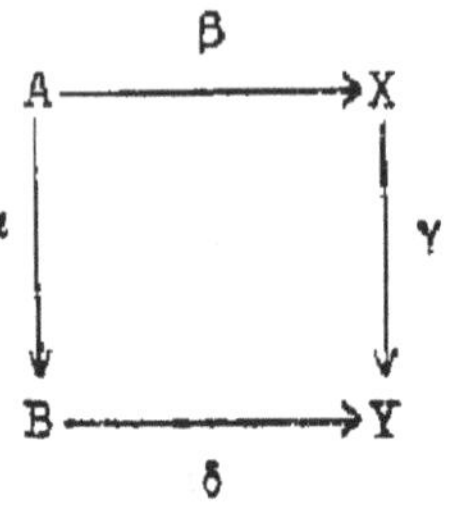

there is a unique map $A \to B\times_Y X$ denoted $(\alpha,\beta)_Y$ or simply (α,β) such that $pr_1(\alpha,\beta) = \alpha$ and $pr_2(\alpha,\beta) = \beta$ where $pr_1: B\times_Y X \to B$ and $pr_2: B\times_Y X \to X$ are the canonical projections. Also (2) is said to be <u>cartesian</u> if (α,β) is an isomorphism. We shall denote the <u>cofibre product</u> of B and X under A by $B\vee_A X$ and the two canonical maps by $in_1: B \to B\vee_A X$ and $in_2: X \to B\vee_A X$. The unique map $B\vee_A X \xrightarrow{u} Y$ with $u\, in_1 = \delta$ and $u\, in_2 = \gamma$ will be denoted $\delta +_A \gamma$ or simply $\delta + \gamma$, and (2) will be called <u>co-cartesian</u> if $\delta + \gamma$ is an isomorphism. Finally given a map $f: X \to Y$ there is the <u>diagonal</u> map $\Delta_f = (id_X, id_X): X \to X\times_Y X$ and the <u>codiagonal</u> map $\nabla_f = id_Y + id_Y: Y\vee_X Y \to Y$ of f. We write Δ_X (resp. ∇_Y) if $Y = e$ (resp. $X = \emptyset$).

<u>Definition 3</u>: Let $f,g: A \rightrightarrows B$ be maps. We say that f is <u>left-homotopic</u> to g (notation $f \overset{l}{\sim} g$) if there is a diagram of the form

(3)

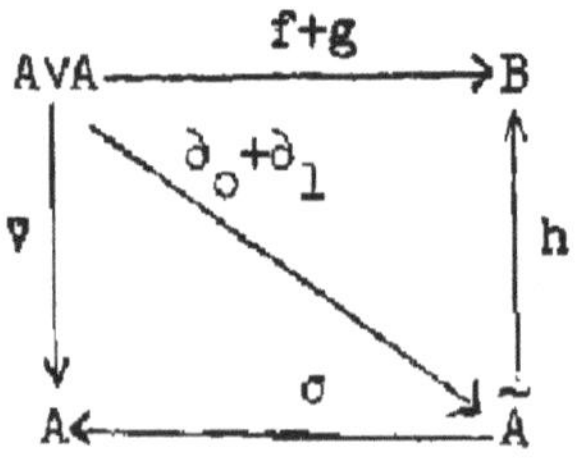

where σ is a weak equivalence. Dually we say that g is <u>right-homotopic</u> to g (notation: $f \overset{r}{\sim} g$) if there is a diagram of the form

(4) 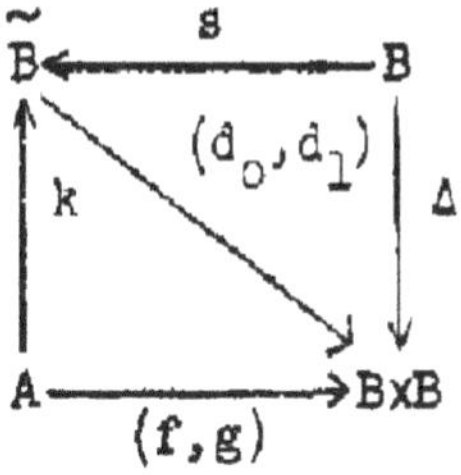

where s is a weak equivalence.

<u>Remark</u>: In example A above two maps of spaces which are homotopic in the usual sense are both left and right homotopic as one sees by taking $\tilde{A} = A\times I$ and $\tilde{B} = B^I$ where I is the unit interval. In fact we have the implications

(5) homotopic $\Longrightarrow$ right homotopic $\Longrightarrow$ left homotopic

where the last implication comes from the dual of Lemma 5(i) below and the fact that every space is fibrant. If A is cofibrant (e.g. a CW complex) then the three notions coincide, but in general it seems that the implications (5) are strict.

<u>Definition 4</u>: By <u>cylinder object</u> for an object A of we mean an object $A\times I$ together with maps $A \vee A \xrightarrow{\partial_0+\partial_1} A\times I \xrightarrow{\sigma} A$ with $\sigma(\partial_0 + \partial_1) = \nabla_A$ such that $\partial_0 + \partial_1$ is a cofibration and σ is a weak equivalence. Dually, a <u>path object</u> for B shall be an object B^I together with a factorization $B \xrightarrow{s} B^I \xrightarrow{(d_0,d_1)} B\times B$ of Δ_B where s is a weak equivalence and (d_0,d_1) is a fibration. By a <u>left homotopy</u> from

$f:A \to B$ to $g:A \to B$ we mean a diagram (3) where $\partial_0 + \partial_1$ is a cofibration and hence $\tilde{A}$ is a cylinder object for A. Similarly a <u>right homotopy</u> from f to g is a diagram (4) where $\tilde{B}$ is a path object for B.

<u>Remarks</u>: 1. $A \times I$ is <u>not</u> the product of A and an object I nor is it a functor of A. In example 1, the product of a space A and the unit interval is <u>not</u> necessarily a cylinder object of A unless A is cofibrant.

2. Since the dual of a model category is again a model category in an evident way there is a corresponding dual assertion for every assertion we make. In the following we will often give only one form and leave the formulation of the dual assertion to the reader.

<u>Lemma 1</u>: If $f,g \in \mathrm{Hom}(A,B)$ and $f \overset{l}{\sim} g$, then there is a left homotopy $h:A \times I \to B$ from f to g.

<u>Proof</u>: Given diagram (3) use M2 to factor $\partial_0 + \partial_1$ into $A \vee A \xrightarrow{\partial_0' + \partial_1'} A' \xrightarrow{\rho} \tilde{A}$ where $\partial_0' + \partial_1'$ is a trivial cofibration and ρ is a trivial fibration. By M5 $\sigma' = \sigma\rho : A' \to A$ is a weak equivalence so A' with ∂_0', ∂_1', and σ' is a cylinder object for A. $h' = h\rho : A' \to B$ is the desired left homotopy from f to g.

<u>Lemma 2</u>: Let A be a cofibrant object and let $A \times I$ be a cylinder object for A. Then $\partial_0 : A \to A \times I$ and $\partial_1 : A \to A \times I$ are trivial cofibrations.

<u>Proof</u>: $in_1 : A \to A \vee A$ is a cofibration by the cobase change

assertion in M3, hence $\partial_o = (\partial_o + \partial_1)in_1$ is a cofibration. $\sigma\partial_o = id_A$ and M5 imply that ∂_o is also a weak equivalence. Similarly ∂_1 is a trivial cofibration.

Corollary: (Covering Homotopy theorem). Let A be cofibrant and let $p:X \to Y$ be a fibration, let $\alpha:A \to X$, and let $h:A\times I \to Y$ be a left homotopy with $h\partial_o = p\alpha$. Then there is a left homotopy $H:A\times I \to X$ with $H\partial_o = \alpha$ and $pH = h$.

Proof: By M1, H exists in

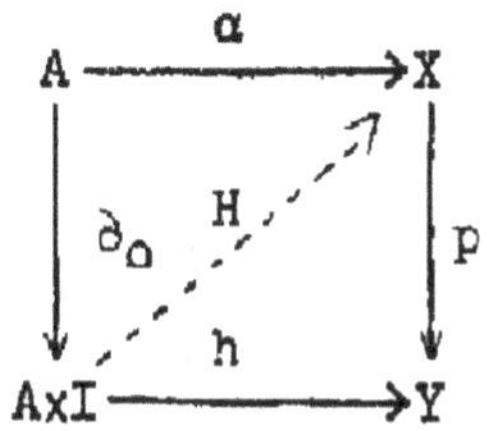

The dual assertion is the homotopy extension theorem.

Lemma 3: Let A be cofibrant and let $A\times I$ and $A\times I'$ be two cylinder objects for A. Then the result of "gluing" $A\times I$ to $A\times I'$ by the identification $\partial_1 A = \partial_o' A$, defined precisely to be the object $\tilde{A}$ in the co-Cartesian diagram

(6)

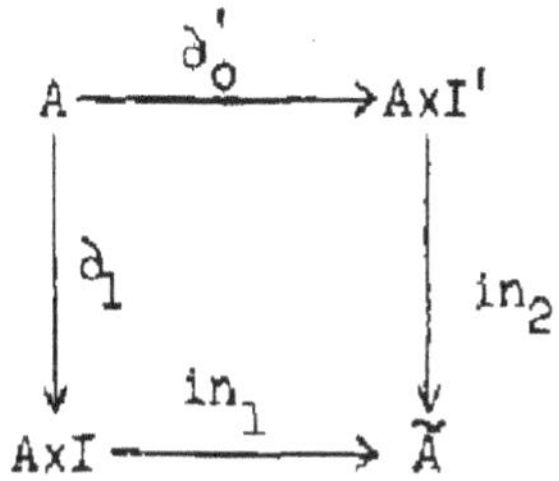

is also a cylinder object $A\times I''$ for A with $\partial_o'' = in_1\partial_o, \partial_1'' = in_2\partial_1'$, $\sigma'' in_1 = \sigma$, and $\sigma'' in_2 = \sigma'$.

Proof: M4 and Lemma 2 show that in_1 and in_2 are weak equivalences; as $\partial''_o = in_1\partial_o$, $\sigma''\partial''_o = id_A$ we have by M5 that $\sigma'':\tilde{A} \to A$ is a weak equivalence. $\partial''_o + \partial''_1 : A \vee A \to \tilde{A}$ is the composition of $A \vee A \xrightarrow{in_1 \vee id_A} (A \times I) \vee A$, which is the co-base extension of ∂_o by $A \xrightarrow{in_1} A \vee A_1$, and the map $(A \times I) \vee A \xrightarrow{in_1 + \partial''_o} \tilde{A}$, which is the co-base extension of $\partial'_o + \partial'_1$ by $A \vee A \xrightarrow{\partial_1 + id_A} (A \times I) \vee A$. By M3 $\partial''_o + \partial''_1$ is a cofibration and hence $\tilde{A}$ is a cylinder object for A.

Lemma 4: If A is cofibrant, then $\overset{\ell}{\sim}$ is an equivalence relation on $Hom(A,B)$.

Proof: The relation is reflexive since if $f = g$ we may take $\tilde{A} = A$ and $h = f$ in (3) and it is symmetric since given (3) we may interchange ∂_o and ∂_1. Finally given $f_o, f_1, f_2 \in Hom(A,B)$ and a left homotopy $h: A \times I \to B$ from f_o to f_1 and a left homotopy $h': A \times I' \to B$ we obtain by Lemma 3 a left homotopy $h'': A \times I'' \to B$ from f_o to f_2 by setting $h'' in_1 = h$ and $h'' in_2 = h'$.

Lemma 5: Let A be cofibrant and let $f, g \in Hom(A,B)$. Then

(i) $f \overset{\ell}{\sim} g \Longrightarrow f \overset{r}{\sim} g$

(ii) $f \overset{r}{\sim} g \Longrightarrow$ there exists a right homotopy $k: A \to B^I$ from f to g with $s: B \to B^I$ a trivial cofibration.

(iii) If $u: B \to C$, then $f \overset{r}{\sim} g \Longrightarrow uf \overset{r}{\sim} ug$.

Proof: (i) By Lemma 1 there is a left homotopy $h: A \times I \to B$ from f to g and by M2 there is a path object B^I for B.

By Lemma 2 and M1 the dotted arrow K exists in

(7)

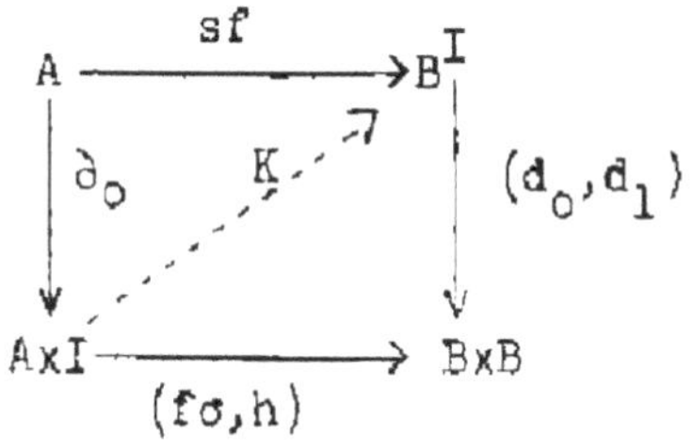

and $k = K\partial_1 : A \to B^I$ is the desired right-homotopy from f to g

(ii) Let $k' : A \to B^I$ be a right homotopy from f to g and let $B \xrightarrow{s} \tilde{B} \xrightarrow{\rho} B^{I'}$ be a factorization of $s' : B \to B^{I'}$, into a trivial cofibration followed by a fibration. By M5 ρ is a weak equivalence. Let $(d_0, d_1) = (d_0', d_1')\rho : \tilde{B} \to B \times B$ so that (d_0, d_1) is a fibration by M3 and hence $\tilde{B}$ with d_0, d_1, and s is a path object for B. By M1 there is a dotted arrow k in

(8)

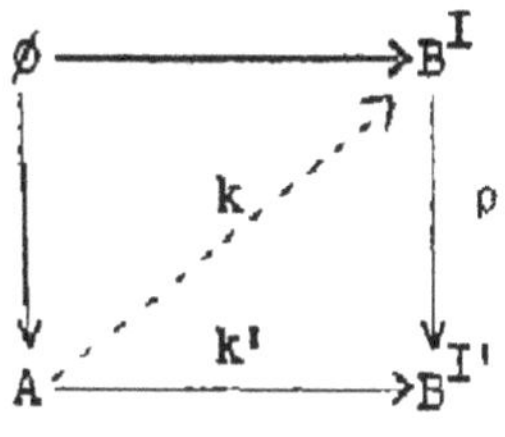

and k gives the desired right homotopy from f to g.

(iii) Let k be as in (ii) and let C^I be a path object for C. By M1 it is possible to lift in

(9)

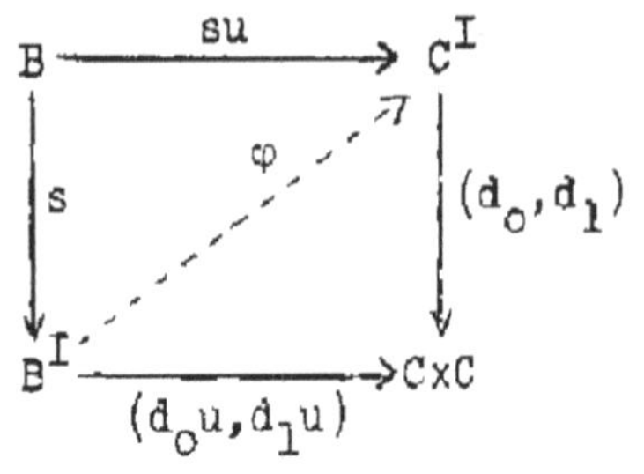

and $k\varphi : A \to C^I$ is a right homotopy from uf to ug. Q.E.D.

If A and B are objects of $\mathcal{C}$ we let $\pi^r(A,B)$ (resp. $\pi^\ell(A,B)$) be the set of equivalence classes of $\mathrm{Hom}(A,B)$ with respect to the equivalence relation generated by $\underset{\sim}{r}$ (resp. $\underset{\sim}{\ell}$). When A cofibrant and B is fibrant, in which case $\underset{\sim}{\ell}$ and $\underset{\sim}{r}$ coincide and are already equivalence relations by Lemmas 4, 5(i) and their duals, we shall denote the relation by $\sim$, call it homotopy and let $\pi_o(A,B)$ or simply $\pi(A,B)$ be the set of equivalence classes.

<u>Lemma 6</u>: If A is cofibrant, then composition in $\mathcal{C}$ induces a map $\pi^r(A,B) \times \pi^r(B,C) \to \pi^r(A,B)$.

<u>Proof</u>: It suffices to show that if $f,g \in \mathrm{Hom}(A,B)$, $u \in \mathrm{Hom}(B,C)$ and $f \underset{\sim}{r} g$ then $uf \underset{\sim}{r} ug$, which is Lemma 5(iii), and that if $u,v \in \mathrm{Hom}(B,C)$, $f \in \mathrm{Hom}(A,B)$, and $u \underset{\sim}{r} v$, then $uf \underset{\sim}{r} vf$, which is immediate from the definition.

<u>Lemma 7</u>: Let A be cofibrant and let $p : X \to Y$ be a trivial fibration. Then p induces a bijection $p_* : \pi^\ell(A,X) \xrightarrow{\sim} \pi^\ell(A,Y)$.

<u>Proof</u>: The map is well-defined since $f \underset{\sim}{\ell} g \Longrightarrow pf \underset{\sim}{\ell} pg$ is immediate from the definition. The map is surjective by M1.

By Lemma 4 if $f, g \in \mathrm{Hom}(A,X)$ and pf, pg represent the same element of $\pi^{\ell}(A,Y)$, then there is a left homotopy $h: A \times I \to Y$ from pf to pg. If H is a lifting in

(10)

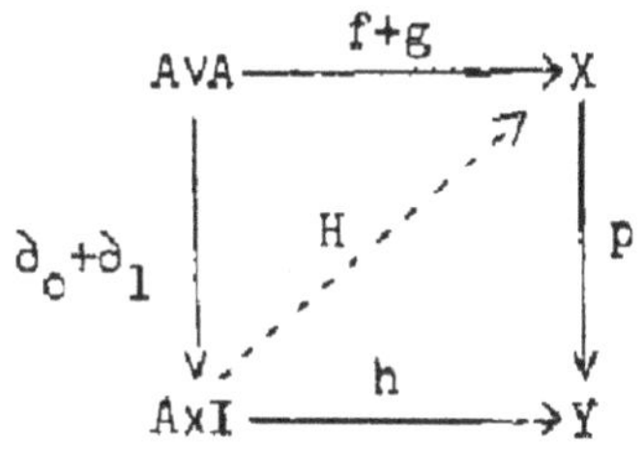

then H is a left homotopy from f to g. This shows that p_* is injective.

Let $\mathcal{C}_c$, $\mathcal{C}_f$, and $\mathcal{C}_{cf}$ be the full subcategories of consisting of the cofibrant, fibrant, and both fibrant and cofibrant objects of $\mathcal{C}$ respectively. By Lemma 6 we may define a category $\pi\mathcal{C}_c$ with the same objects as $\mathcal{C}_c$, with $\mathrm{Hom}_{\pi\mathcal{C}_c}(A,B) = \pi^r(A,B)$ and with composition induced from that of $\mathcal{C}$. If we denote the right homotopy class of a map $f: A \to B$ by $\bar{f}$ we obtain a functor $\mathcal{C}_c \to \pi\mathcal{C}_c$ given by $X \to X$, $f \to \bar{f}$. Similarly by the dual of Lemma 6 we may define $\pi\mathcal{C}_f$ $\left(\text{resp. } \pi\mathcal{C}_{cf}\right)$ to be the category with the same objects as $\mathcal{C}_f$ and with $\pi^{\ell}(A,B)$ (resp. $\pi(A,B)$) as maps from A to B.

<u>Definition 5</u>: Let $\mathcal{C}$ be an arbitrary category and let S be a subclass of the class of maps of $\mathcal{C}$. By the <u>localization</u> of $\mathcal{C}$ with respect to S we mean a category $S^{-1}\mathcal{C}$ together with a functor $\gamma: \mathcal{C} \to S^{-1}\mathcal{C}$ having the following universal property. For every $s \in S$, $\gamma(s)$ is an isomorphism; given any

functor $F: \mathcal{C} \to \mathcal{B}$ with $F(s)$ an isomorphism for all $s \in S$, there is a unique functor $\theta: S^{-1}\mathcal{C} \to \mathcal{B}$ such that $\theta \circ \gamma = F$.

Except for set-theoretic difficulties the category $S^{-1}\mathcal{C}$ exists and may be constructed by mimicking the construction of the free group (see Gabriel-Zisman [7]).

<u>Definition 6</u>: Let $\mathcal{C}$ be a model category. Then the <u>homotopy category</u> of $\mathcal{C}$ is the localization of $\mathcal{C}$ with respect to the class of weak equivalences and is denoted by $\gamma: \mathcal{C} \to \mathrm{Ho}\,\mathcal{C}$. $\gamma_c: \mathcal{C}_c \to \mathrm{Ho}\,\mathcal{C}_c$ (resp. $\gamma_f: \mathcal{C}_f \to \mathrm{Ho}\,\mathcal{C}_f$) will denote the localization of $\mathcal{C}_c$ with respect to the class of maps in $\mathcal{C}_c$ (resp. $\mathcal{C}_f$) which are weak equivalences in $\mathcal{C}$. We sometimes use the notation $[X,Y]$ for $\mathrm{Hom}_{\mathrm{Ho}\,\mathcal{C}}(X,Y)$.

<u>Lemma 8</u>: (i) Let $F: \mathcal{C} \to \mathcal{B}$ carry weak equivalences in $\mathcal{C}$ into isomorphisms in $\mathcal{B}$. If $f \overset{l}{\sim} g$ or $f \overset{r}{\sim} g$, then $F(f) = F(g)$ in $\mathcal{B}$.

(ii) Let $F: \mathcal{C}_c \to \mathcal{B}$ carry weak equivalences in $\mathcal{C}_c$ into isomorphisms in $\mathcal{B}$. If $f \overset{r}{\sim} g$, then $F(f) = F(g)$ in $\mathcal{B}$.

<u>Proof</u>: (i) Let $h: A \times I \to B$ be a left homotopy from f to g. As σ is a weak equivalence, $F(\sigma)$ is an isomorphism. As $F(\sigma)F(\partial_0) = F(\sigma)F(\partial_1) = \mathrm{id}_A$, $F(\partial_0) = F(\partial_1)$ and so $F(f) = F(h)F(\partial_0) = F(h)F(\partial_1) = F(g)$.

(ii) The proof is the same as (i) since by Lemma 4 (ii) we may assume that $s: B \to B^I$ is a cofibration and hence B^I is in $\mathcal{C}_c$.

By Lemma 8 the functors $\gamma_c, \gamma_f, \gamma$ induce functors

$\bar{\gamma}_c : \pi\mathcal{C}_c \to \mathrm{Ho}\,\mathcal{C}_c$, $\bar{\gamma}_f : \pi\mathcal{C}_f \to \mathrm{Ho}\,\mathcal{C}_f$, and $\bar{\gamma} : \pi\mathcal{C}_{cf} \to \mathrm{Ho}\,\mathcal{C}$, provided these localizations exist. The following result shows that the homotopy category $\mathrm{Ho}\,\mathcal{C}$ as defined in Definition 6 is equivalent to the more concrete category $\pi\mathcal{C}_{cf}$.

<u>Theorem 1'</u>: $\mathrm{Ho}\,\mathcal{C}$ exist and the functor $\bar{\gamma} : \pi\mathcal{C}_{cf} \to \mathrm{Ho}\,\mathcal{C}$ is an equivalence of categories.

This is included in the following more complex assertion which is presented for the purpose of comparison with (Gabriel-Zisman [7]).

<u>Theorem 1</u>: The categories $\mathrm{Ho}\,\mathcal{C}$, $\mathrm{Ho}\,\mathcal{C}_c$, $\mathrm{Ho}\,\mathcal{C}_f$ exist and there is a diagram of functors

$$
\begin{array}{ccc}
\pi\mathcal{C}_c & \xrightarrow{\ \bar{\gamma}_c\ } & \mathrm{Ho}\,\mathcal{C}_c \\
\uparrow\rlap{\scriptstyle\cup} & & \downarrow\wr \\
\pi\mathcal{C}_{cf} & \xrightarrow[\sim]{\ \bar{\gamma}\ } & \mathrm{Ho}\,\mathcal{C} \\
\downarrow\rlap{\scriptstyle\cap} & & \uparrow\wr \\
\pi\mathcal{C}_f & \xrightarrow{\ \bar{\gamma}_f\ } & \mathrm{Ho}\,\mathcal{C}_f
\end{array}
\tag{11}
$$

where $\hookrightarrow$ denotes a full embedding and $\xrightarrow{\sim}$ denotes an equivalence of categories. Furthermore if $(\bar{\gamma})^{-1}$ is a quasi-inverse for $\bar{\gamma}$, then the fully faithful functor $\mathrm{Ho}\,\mathcal{C}_c \xrightarrow{\sim} \mathrm{Ho}\,\mathcal{C} \xrightarrow[\sim]{(\bar{\gamma})^{-1}} \pi\mathcal{C}_{cf} \hookrightarrow \pi\mathcal{C}_c$ is right adjoint to $\bar{\gamma}_c$ and the fully faithful functor $\mathrm{Ho}\,\mathcal{C}_f \xrightarrow{\sim} \mathrm{Ho}\,\mathcal{C} \xrightarrow[\sim]{(\bar{\gamma})^{-1}} \pi\mathcal{C}_{cf} \hookrightarrow \pi\mathcal{C}_f$ is left adjoint to $\bar{\gamma}_f$.

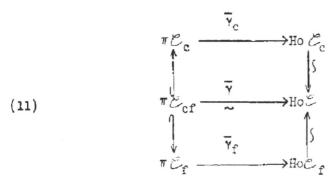

Proof: For each object X choose a trivial fibration $p_X:Q(X) \to X$ with $Q(X)$ cofibrant and a trivial fibration $i_X:X \to R(X)$ with $R(X)$ fibrant. We assume that $Q(X) = X$ and $p_X = id_X$ (resp. $X = R(X)$ and $i_X = id_X$) if X is already cofibrant (resp. fibrant). For each map $f:X \to Y$ we may choose by M1 a map $Q(f):Q(X) \to Q(Y)$ (resp. $R(f)i_X = i_Y f$) which is unique up to left (resp. right) homotopy by Lemma 7. It follows that $Q(gf) \overset{\ell}{\sim} Q(g)Q(f)$ and $Q(id_X) \overset{\ell}{\sim} id_{Q(X)}$, hence $Q(gf) \overset{r}{\sim} Q(g)Q(f)$ and $Q(id_X) \overset{r}{\sim} id_{Q(X)}$ by Lemma 4(i), and therefore $X \to Q(X)$ $f \to \overline{Q(f)}$ is a well-defined functor which we shall denote $\overline{Q}:\mathcal{C} \to \pi\mathcal{C}_c$. Similarly there is a functor $\overline{R}:\mathcal{C} \to \pi\mathcal{C}_f$.

If X is cofibrant, $f,g \in \mathrm{Hom}(X,Y)$, and $f \overset{r}{\sim} g$, then by Lemma 4(iii) $i_Y f \overset{r}{\sim} i_Y g$ and hence $R(f) \sim R(g)$ by the dual of Lemma 7. It follows that $\overline{R}$ restricted to $\mathcal{C}_c$ induces a functor $\pi\mathcal{C}_c \to \pi\mathcal{C}_{cf}$ and that there is a well-defined functor $\overline{RQ}:\mathcal{C} \to \pi\mathcal{C}_{cf}$ given by $X \to RQX$, $f \to \overline{RQ(f)}$.

Let $\mathrm{Ho}\mathcal{C}$ be the category having the same objects as $\mathcal{C}$ with $\mathrm{Hom}_{\mathrm{Ho}\mathcal{C}}(X,Y) = \mathrm{Hom}_{\pi\mathcal{C}_{cf}}(RQX,RQY) = \pi(RQX,RQY)$ and the obvious composition. Let $\gamma:\mathcal{C} \to \mathrm{Ho}\mathcal{C}$ be given by $\gamma(X) = X$, $\gamma(f) = \overline{RQ(f)}$. As $RQ(X) = X$ if X is in $\mathcal{C}_{cf}$, it is clear that the functor $\overline{\gamma}:\pi\mathcal{C}_{cf} \to \mathrm{Ho}\mathcal{C}$ induced by γ is fully faithful. By Lemma 7 and its dual, trivial fibrations and trivial cofibration in $\mathcal{C}_{cf}$ become isomorphisms in $\pi\mathcal{C}_{cf}$; hence any weak equivalence in $\mathcal{C}_{cf}$ becomes an isomorphism in $\pi\mathcal{C}_{cf}$ by M2 and M5. If $f:X \to Y$ is a weak equivalence in $\mathcal{C}$, then

$fp_X = p_Y Q(f)$ and M5 imply that $Q(f)$ is a weak equivalence in $\mathcal{C}_c$ and similarly $RQ(f)$ is a weak equivalence in $\mathcal{C}_{cf}$; hence $\gamma(f) = \overline{RQ(f)}$ is an isomorphism. It follows that for any X the maps $X \xleftarrow{p_X} Q(X) \xrightarrow{i_{Q(X)}} RQX$ yield an isomorphism of X and $RQ(X)$ in $\text{Ho}\,\mathcal{C}$ and hence $\pi\mathcal{C}_{cf} \xrightarrow{\bar{\gamma}} \text{Ho}\,\mathcal{C}$ is an equivalence of categories.

We now show that $\gamma: \mathcal{C} \to \text{Ho}\,\mathcal{C}$ has the required universal property of Definition 5. As mentioned above γ carries weak equivalences in $\mathcal{C}$ into isomorphisms in $\text{Ho}\,\mathcal{C}$. Let $F: \mathcal{C} \to \mathcal{B}$ do the same. Define $\theta: \text{Ho}\,\mathcal{C} \to \mathcal{B}$ by $\theta(X) = F(X)$ and for $\alpha \in \text{Hom}_{\text{Ho}\,\mathcal{C}}(X,Y)$ choose $f: RQ(X) \to RQ(Y)$ representing α and let $\theta(\alpha)$ be given by the diagram

(12)
$$\begin{array}{ccc}
F(X) & \xrightarrow{\theta(\alpha)} & F(Y) \\
{\scriptstyle F(p_X)}\uparrow\wr & & \uparrow\wr{\scriptstyle F(p_Y)} \\
F(QX) & & F(QY) \\
{\scriptstyle F(i_{QX})}\downarrow\wr & & \downarrow\wr{\scriptstyle F(i_{QY})} \\
F(RQX) & \xrightarrow{F(f)} & F(RQY)
\end{array}$$

By Lemma 8(i) $\theta(\alpha)$ is independent of the choice of f and it is then clear that θ is a functor, in fact the unique functor with $\theta \circ \gamma = F$. This proves the existence of $\text{Ho}\,\mathcal{C}$ and also the horizontal equivalence in (11).

The existence of $\text{Ho}\,\mathcal{C}_c$ and the equivalence $\pi\mathcal{C}_{cf} \xrightarrow{\sim} \text{Ho}\,\mathcal{C}_c$

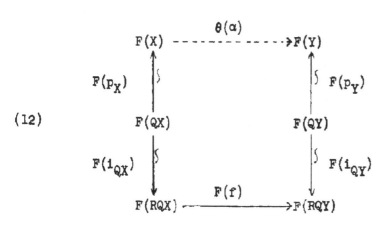

can be proved in the same way using the functor $\mathcal{C}_c \to \pi\mathcal{C}_{cf}$ induced by $\bar{R}$ and Lemma 8(ii). The last assertion of the theorem results from the fact that the inclusion functor $\pi\mathcal{C}_{cf} \hookrightarrow \pi\mathcal{C}_c$ is right adjoint to the functor $\bar{R}' : \pi\mathcal{C}_c \to \pi\mathcal{C}_{cf}$, since $\pi^r(X,Y) \simeq \pi(RX,Y)$ if X is in $\mathcal{C}_c$ and Y is in $\mathcal{C}_{cf}$ by Lemma 7, and from the fact that up to the equivalence $\mathrm{Ho}\,\mathcal{C}_c \simeq \mathrm{Ho}\,\mathcal{C} \simeq \pi\mathcal{C}_{cf}$, $\bar{\gamma}_c : \mathcal{C}_c \to \mathrm{Ho}\,\mathcal{C}_c$ "is" the functor $\bar{R}'$. Q.E.D.

<u>Corollary 1</u>: If A is cofibrant and B is fibrant, then

$$\mathrm{Hom}_{\mathrm{Ho}\mathcal{C}}(A,B) = \pi(A,B)$$

<u>Proof</u>: $\mathrm{Hom}_{\mathrm{Ho}\mathcal{C}}(A,B) = \pi(RQA,RQB) = \pi(RA,QB) \simeq \pi(A,QB) \simeq \pi(A,B)$ by Lemma 7 and its dual.

<u>Corollary 2</u>: The functor $\bar{\gamma}_c : \pi\mathcal{C}_c \to \mathrm{Ho}\,\mathcal{C}_c$ permits calculations by left fractions and the functor $\bar{\gamma}_f : \pi\mathcal{C}_f \to \mathrm{Ho}\,\mathcal{C}_f$ permits calculation by right fractions.

<u>Proof</u>: This follows from the first chapter of (Gabriel-Zisman), since $\bar{\gamma}_c$ has a fully faithful right adjoint.

<u>Remarks</u>: 1. In general the localization $\mathcal{C} \to \mathrm{Ho}\,\mathcal{C}$ cannot be calculated by either left or right fractions.

2. In example A, $\mathcal{C} = \mathcal{C}_f$ and the usual homotopy relation on maps coincides with homotopy in the sense of Definition 5 on $\mathcal{C}_c$. Thus $\pi\mathcal{C}_{cf} = \pi\mathcal{C}_c$ is the homotopy category of cofibrant spaces which in turn is equivalent to the usual homotopy category of CW complexes. In example B, $\mathcal{C} = \mathcal{C}_f$ and homotopy on $\mathcal{C}_c$ coincides with the chain homotopy relation.

Hence $\pi\mathcal{C}_c = \pi\mathcal{C}_{cf}$ is what is denoted $K^-(\underline{P})$ in Harshorne [9]) where $\underline{P}$ is the additive sub-category of projectives in $\mathcal{Q}$, while $\mathrm{Ho}\,\mathcal{C}$ is the derived category $D^-(\mathcal{Q})$ or $D_+(\mathcal{Q})$.

3. The following example shows that although $\mathrm{Ho}\,\mathcal{C}$ is determined by the category $\mathcal{C}$ and the class of weak equivalences, the model structure on $\mathcal{C}$ isn't. Let $\mathcal{Q}$ be an abelian category of finite homological dimension having enough projectives and injectives. Then $\mathcal{C} = C_b(\mathcal{Q})$ the category of bounded complexes is what one should call a full sub-model category of $C_+(\mathcal{Q})$ as in example B. The dual of example B gives the structure of a model category on $C_-(\mathcal{Q})$, the category of complexes bounded above, where cofibrations are injections, fibrations are surjective maps with injective kernels, and weak equivalences are homology isomorphisms. Again $C_b(\mathcal{Q})$ is a full-sub-model category of $C_-(\mathcal{Q})$ and we obtain different model structures on $C_b(\mathcal{Q})$ with the same family of weak equivalences.

§2. The loop and suspension functors

Homotopy theory is concerned not only with the category $\mathrm{Ho}\,\mathcal{C}$ as a category but also with certain extra structure which comes from performing constructions in $\mathcal{C}$. In this section we will be concerned with one aspect of this extra structure--the loop and suspension functors.

$\mathcal{C}$ denotes a fixed model category and $f,g:A \rightrightarrows B$ two maps in $\mathcal{C}$ where A is cofibrant and B is fibrant.

Definition 1: Let $h:A\times I \to B$ and $h':A\times I' \to B$ be two left homotopies from f to g. By a left homotopy from h to h' we mean a diagram

(1)

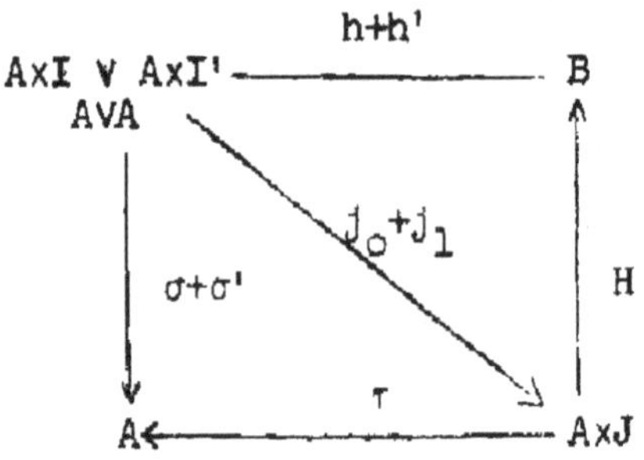

where $j_0 + j_1$ is a cofibration and τ is a weak equivalence. (Here $A\times I \underset{A\vee A}{\vee} A\times I'$ is the cofibre product of the maps $\partial_0 + \partial_1 : A\vee A \to A\times I$ and $\partial_0' + \partial_1' : A\vee A \to A\times I'$.) We say h is left homotopic to h' (notation $h \overset{\ell}{\sim} h'$) if such a left homotopy exists.

Remarks: 1. As in §1, the symbol $A\times J$ will denote an object of $\mathcal{C}$ together with a cofibration $j_0 + j_1$ and weak equivalence τ as in (1). $A \times J$ is <u>not</u> generally the product of A and an object "J".

2. There is a dual notion of right homotopy of right homotopies whose formulation we leave to the reader.

<u>Definition 2</u>: Let $h: A \times I \to B$ be a left homotopy from f to g and let $k: A \to B^I$ be a right homotopy from f to g. By a <u>correspondence</u> between h and k we mean a map $H: A \times I \to B^I$ such that $H\partial_0 = k$, $H\partial_1 = sg$, $d_0 H = h$, and $d_1 H = g\sigma$. We say that h and k <u>correspond</u> if such a correspondence exists.

It will be useful to use the following diagrams to indicate a left homotopy h, a right homotopy k, and a correspondence H between h and k

(2)

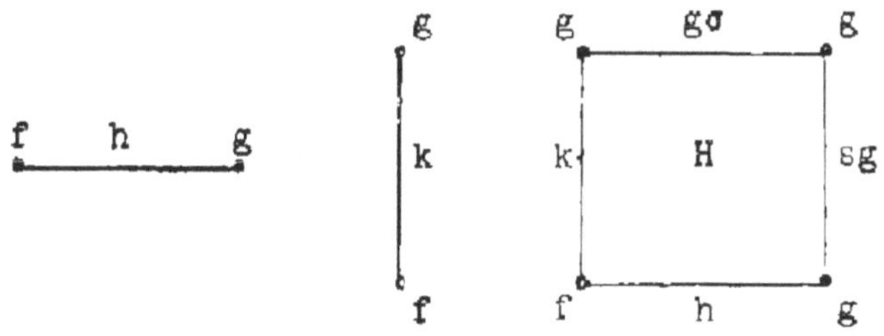

respectively.

<u>Lemma 1</u>: Given $A \times I$ and a right homotopy $k: A \to B^I$ there is a left homotopy $h: A \times I \to B$ corresponding to k. Dually given B^I and h, there is a k corresponding to h.

<u>Proof</u>: Same as that of Lemma 5(ii), §1.

<u>Lemma 2</u>: Suppose that $h: A \times I \to B$ and $h': A \times I' \to B$ are two left homotopies from f to g and that $k: A \to B^I$ is a right homotopy from f to g. Suppose that h and k correspond.

Then h' and k correspond iff h' is left homotopic to h.

<u>Proof</u>: Let $H: A \times I \to B^I$ be a correspondence between h and k, and let $H': A \times I \to B^I$ be a correspondence between h' and k. Let $A \times J$, $j_0 + j_1$, and τ be as in Remark 1. The dotted arrow K exists in

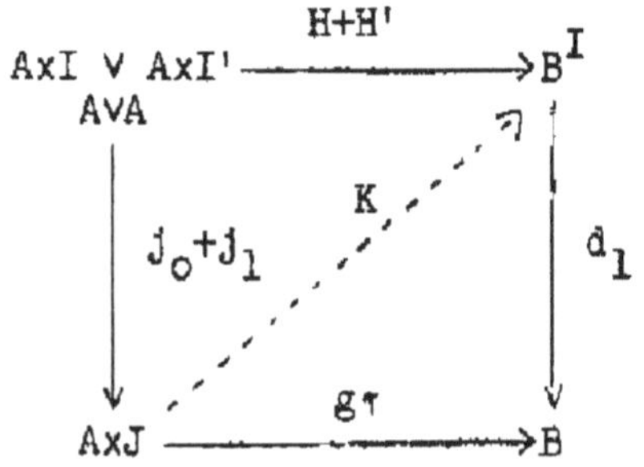

and $d_0 K: A \times J \to B$ is a left homotopy from h to h'. Conversely suppose given $H: A \times I \to B^I$ and a left homotopy $K: A \times J \to B$ from h to h'. Then $j_0: A \times I \to A \times J$ is a cofibration by M3 since it's the composition of $j_0 + j_1$ and $in_1: A \times I \to A \times I \vee_{A \vee A} A \times I'$ which is the cobase extension of $\partial_0 + \partial_1$. Also j_0 is trivial by M5 since $\tau j_0 = \sigma$. Hence the dotted arrow φ exists in

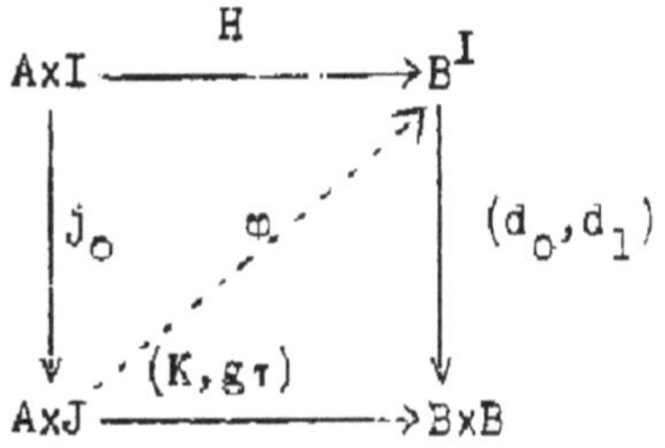

and $\varphi j_1: A \times I' \to B^I$ is a correspondence between h' and k. Q.E.D.

<u>Corollary</u>: "is left homotopic to" is an equivalence relation on the class of left homotopies from f to g and the equivalence classes form a set $\pi_1^{\ell}(A,B;f,g)$. Dually right homotopy classes of right homotopies form a set $\pi_1^{r}(A,B;f,g)$. Correspondence yields a bijection $\pi_1^{\ell}(A,B;f,g) \simeq \pi_1^{r}(A,B;f,g)$.

<u>Proof</u>: Lemma 2 yields the equivalence relation assertion while Lemma 1 shows that every h is equivalent to a $k:A \to B^I$ with <u>fixed</u> B^I and hence the equivalence classes form a set. The last assertion is clear from Lemma 2 and its dual.

By the corollary we may drop the "ℓ" and "r" and write $\pi_1(A,B;f,g)$ and refer to an element of this set as a homotopy class of homotopies from f to g.

Again let $\mathcal{C}$ be a fixed model category, let A be a cofibrant object of $\mathcal{C}$, and let B be a fibrant object.

<u>Definition 3</u>: Let $f_1,f_2,f_3 \in \mathrm{Hom}(A,B)$, let $h:A\times I \to B$ be a left homotopy from f_1 to f_2 and let $h':A\times I' \to B$ be a left homotopy from f_2 to f_3. By the <u>composition</u> of h and h', denoted $h\cdot h'$, we mean the homotopy $h'':A\times I'' \to B$ given by $h''in_1 = h$, $h''in_2 = h'$ where $A\times I''$ is the path object constructed in Lemma 3, §1. If $f,g \in \mathrm{Hom}(A,B)$ and $h:A\times I \to B$ is a left homotopy from f to g, then by the <u>inverse</u> of h, denoted h^{-1} we mean the left homotopy $h':A\times I' \to B$ from g to f, where $A\times I'$ is the path object for A given by $A\times I' = A\times I$, $\partial_0' = \partial_1$, $\partial_1' = \partial_0$, $\sigma' = \sigma$ and where $h' = h$.

The following pictures for $h\cdot h'$ and h^{-1} will be used

(3)

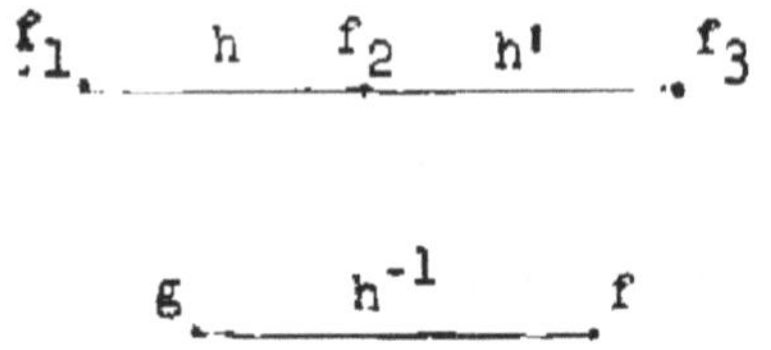

Composition and inverses for right homotopies are defined dually and will be pictured by diagrams like (3) but where the lines run vertically.

<u>Proposition 1</u>: Composition of left homotopies induces maps $\pi_1^{\ell}(A,B;f_1,f_2) \times \pi_1^{\ell}(A,B;f_2,f_3) \to \pi_1^{\ell}(A,B;f_1,f_3)$ and similarly for right homotopies. Composition of left and right homotopies is compatible with the correspondence bijection of the corollary to Lemma 2. Finally the category with objects $Hom(A,B)$, with a morphism from f to g defined to be an element of $\pi_1(A,B;f,g)$, and with composition of morphisms defined to be induced by composition of homotopies, is a groupoid, the inverse of an element of $\pi_1^{\ell}(A,B;f,g)$ represented by h being represented by h^{-1} .

<u>Proof</u>: Let h (resp. k) be a left (resp. right) homotopy from f_1 to f_2 , let h' (resp. k') be a left (resp. right) homotopy from f_2 to f_3, and let H (resp. H') be a correspondence between h and k (resp. h' and k'). Then we have the following correspondence between $h \cdot h'$ and $k \cdot k'$.

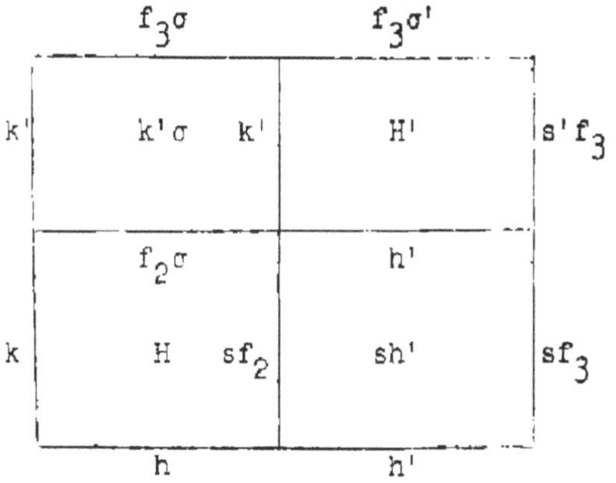

Taking Lemma 2 into account this proves the first two assertions of the proposition.

Composition is associative because $(h\cdot h')\cdot h''$ and $h\cdot(h'\cdot h'')$ are both represented by the picture

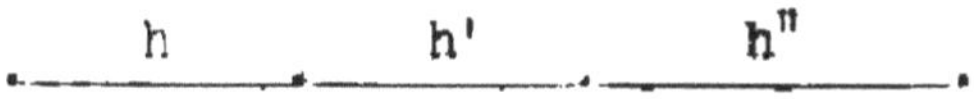

If $h:A\times I \to B$ is a left homotopy from f to g and $H:A\times I \to B^I$ is a correspondence of h with some right homotopy k then the diagrams

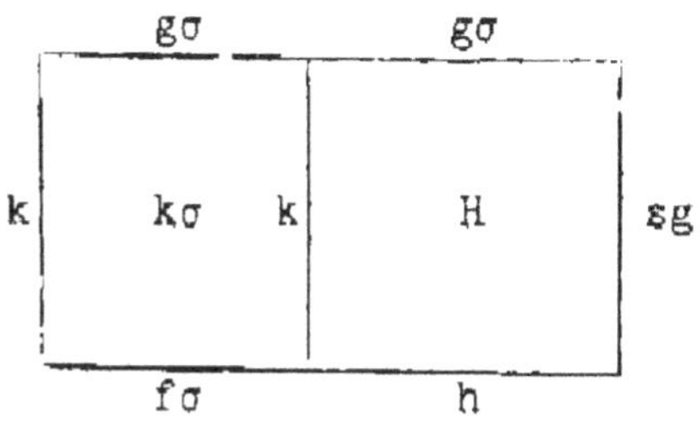

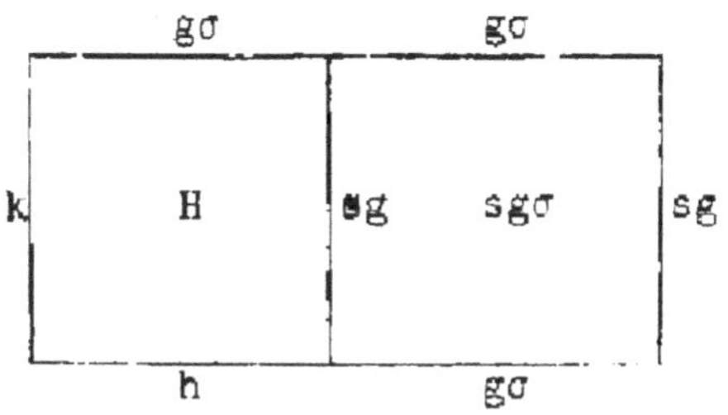

and Lemma 2 give $f\sigma \cdot h \sim h$, $h \cdot g\sigma \sim h$, proving the existence of identities and hence that $\mathrm{Hom}(A,B)$ is a category. Finally let $H':A\times I' \to B^I$ be $H:A\times I \to B^I$, where $A\times I'$ is $A\times I$ with $\partial'_o = \partial_1$, $\partial'_1 = \partial_o$, and $\sigma' = \sigma$, and let $H'':A\times I' \to B^I$ be a correspondence of $h^{-1}:A\times I' \to B$ with some $k'':A \to B^I$, and let $\tilde{H}:A\times I \to B^I$ be H''. Then the diagrams

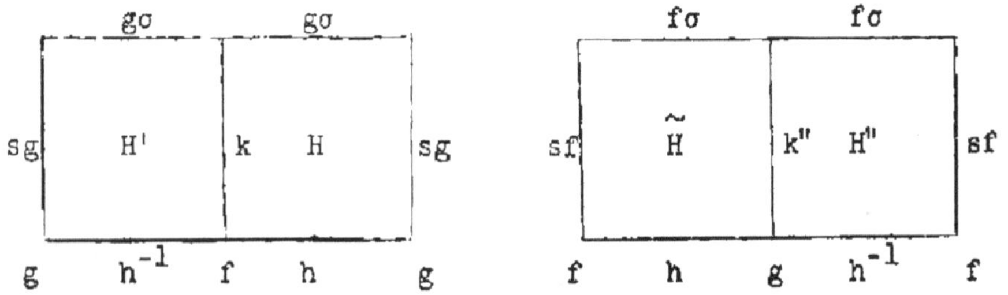

show that $h^{-1}\cdot h \sim g\sigma$ and $h \cdot h^{-1} \sim f\sigma$ proving the last assertion of Proposition 1. Q.E.D.

It is clear that if $i:A' \to A$ is a map of cofibrant objects, then there is a functor $i^*:\mathrm{Hom}(A,B) \to \mathrm{Hom}(A',B)$ which sends f into fi and a right homotopy $k:A \to B^I$ into $ki:A' \to B^I$. Similarly if $j:B \to B'$ is a map of fibrant objects there is a functor $j_*:\mathrm{Hom}(A,B) \to \mathrm{Hom}(A,B')$.

<u>Lemma 3</u>: The diagram

$$\begin{array}{ccc} \pi_1(A,B;f,g) & \xrightarrow{i^*} & \pi_1(A',B;fi,gi) \\ \downarrow j_* & & \downarrow j_* \\ \pi_1(A,B';jf,jg) & \xrightarrow{i^*} & \pi_1(A',B',jfi,jgi) \end{array}$$

commutes.

<u>Proof</u>: Let $\alpha \in \pi_1(A,B;f,g)$ and represent α by $h:A\times B \to B$, $k:A \to B^I$, and let H be a correspondence between h and k. By Lemma 5(ii), §1 and Lemma 1, §2 we may assume that $\sigma:A\times I \to A$ is a trivial fibration and $s:B \to B^I$ is a trivial cofibration. By M1 we may then choose dotted arrows in

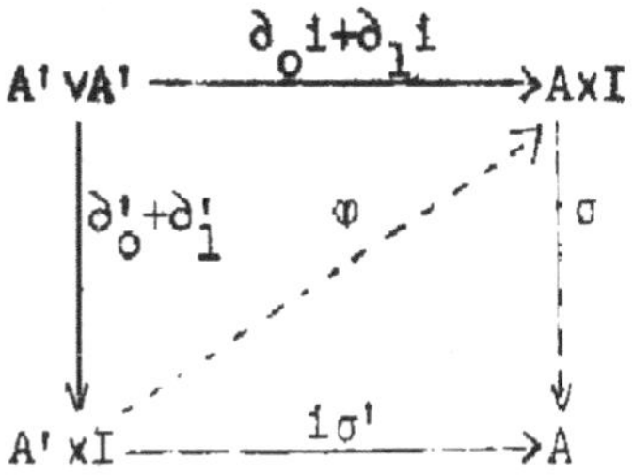

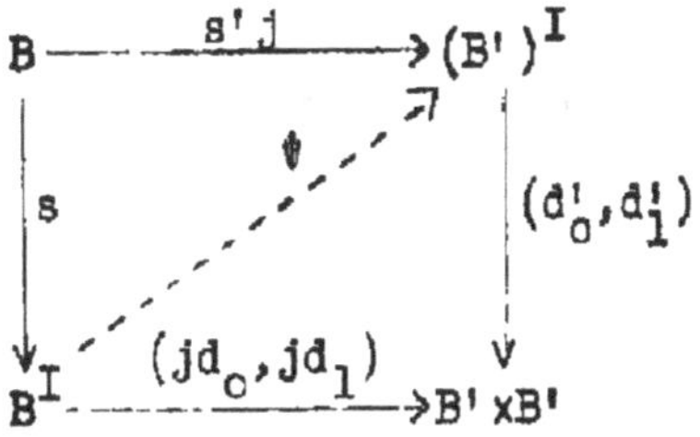

Then H is a correspondence between jh and ψk; hence ψk represents $j_*\alpha$ and so ψki represents $i^*j_*\alpha$. Similarly $H\varphi$ is a correspondence between ki and $h\varphi$; hence $h\varphi$ represents $i^*\alpha$ and so $jh\varphi$ represents $j_*i^*\alpha$. Finally $\psi H\varphi$ is a correspondence between ψki and $jh\varphi$ which shows that $i^*j_*\alpha = j_*i^*\alpha$. Q.E.D.

<u>Definition 4</u>: A <u>pointed category</u> is a category $\mathcal{A}$ in which "the" initial object and final object exist and are isomorphic. We shall denote this object by $*$ and call it the <u>null-object</u> of $\mathcal{A}$. If X and Y are arbitrary objects of $\mathcal{A}$ we denote by $0 \in \mathrm{Hom}_{\mathcal{A}}(X,Y)$ the composition $X \to * \to Y$. If $f:X \to Y$ is a map in $\mathcal{C}$, then we define the <u>fibre</u> of f to be the fibre product $*\times_Y X$ and the <u>cofibre</u> of f to be the cofibre product

$*V_X Y$.

By a <u>pointed model category</u> we mean a model category $\mathcal{C}$ which is also a pointed category. If A is in $\mathcal{C}_c$ and B in $\mathcal{C}_f$, then we will abbreviate $\pi_1(A,B;0,0)$ to $\pi_1(A,B)$. $\pi_1(A,B)$ is a group by the above proposition.

<u>Theorem 2</u>: Let $\mathcal{C}$ be a pointed model category. Then there is a functor $A,B \to [A,B]_1$ from $(\mathrm{Ho}\,\mathcal{C})^o \times \mathrm{Ho}\,\mathcal{C}$ to (groups) which is determined up to canonical isomorphism by $[A,B]_1 = \pi_1(A,B)$ if A is cofibrant and B is fibrant. Furthermore, there are two functors from $\mathrm{Ho}\,\mathcal{C}$ to $\mathrm{Ho}\,\mathcal{C}$, the suspension functor Σ and the loop functor Ω and canonical isomorphisms

$$[\Sigma A,B] \simeq [A,B]_1 \simeq [A,\Omega B]$$

of functors $(\mathrm{Ho}\,\mathcal{C})^o \times (\mathrm{Ho}\,\mathcal{C}) \to$ (sets) where $[X,Y] = \mathrm{Hom}_{\mathrm{Ho}}(X,Y)$.

<u>Proof</u>: Let A be cofibrant; choose a cylinder object $A\times I$ and let $A\times I \xrightarrow{\pi} \Sigma A$ be the cofibre of $\partial_0 + \partial_1 : A\vee A \to A\times I$. By M3 ΣA is cofibrant. We shall define a bijection

$$(4) \qquad \rho : \pi(\Sigma A,B) \xrightarrow{\sim} \pi_1(A,B)$$

which is a natural transformation of functors to (sets) as B runs over $\mathcal{C}_f$. Let $\varphi : \Sigma A \to B$ be a map and let $\rho(\varphi)$ be the element of $\pi_1(A,B)$ represented by $\varphi\pi : A\times I \to B$. If $\varphi, \varphi' \in \mathrm{Hom}(\Sigma A,B)$ and $\varphi \sim \varphi'$, then there is a right homotopy $h : \Sigma A \to B^I$ from φ to φ' . Let $H : A\times I \to B^I$ be a correspondence of $\varphi'\pi$ with some right homotopy k from 0 to 0 and consider

the diagram

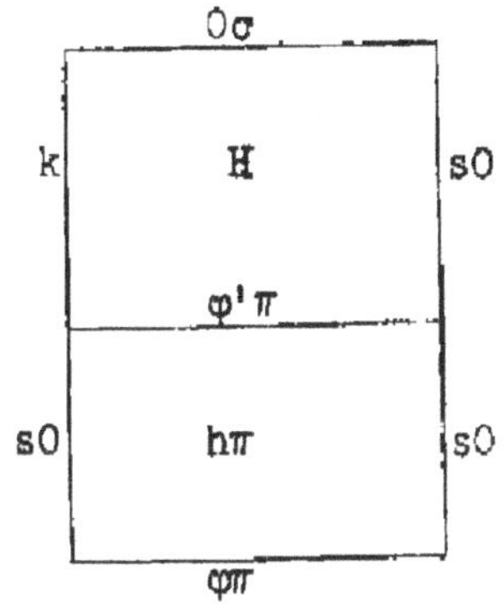

This shows that $\varphi\pi$ corresponds to $s0 \cdot k$ and $\varphi'\pi$ corresponds to k, as $s0 \cdot k$ and k represents the same element of $\pi_1(A,B)$ so do $\varphi\pi$ and $\varphi\pi'$ and hence $\rho(\varphi) = \rho(\varphi')$. This shows that ρ (4) is well-defined. ρ is surjective by Lemma 1. Finally, if $\rho(\varphi) = \rho(\varphi')$, then, with the notation of Definition 1, there is a left homotopy $H: A \times J \to B$ from $\varphi\pi$ to $\varphi'\pi$. Let $H': A \times J \to B$ be given by $H' j_o = H' j_1 = \varphi\pi$ and let K be the dotted arrow in

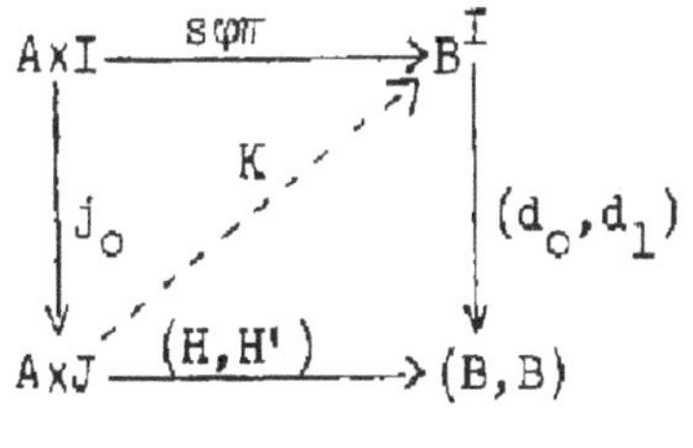

(j_o was shown to be a trivial cofibration in proof of Lemma 2.) Then $Kj_1: A \times I \to B^I$ is a right homotopy from $\varphi\pi$ to $\varphi'\pi$ such that $Kj_1(\partial_o + \partial_1) = 0$ and so induces a right homotopy $\Sigma A \to B^I$

from φ to φ'. This shows ρ is injective and proves (4).

Dually if we choose a path object B^I and let ΩB be the fibre of $(d_0,d_1):B^I \to B\times B$, then ΩB is fibrant and there is a bijection

$$(5) \qquad \pi(A,\Omega B) \xrightarrow{\sim} \pi_1(A,B)$$

which is a natural transformation of functors as A runs over $\mathcal{C}_c$.

Lemma 3 shows that $A,B \to \pi_1(A,B)$ is a functor from $(\mathcal{C}_c)^o \times \mathcal{C}_f$ to (groups). (4) and (5) combined with Theorem 1 and its first corollary show that this functor induces a functor $(\mathrm{Ho}\,\mathcal{C}_c)^o \times \mathrm{Ho}\,\mathcal{C}_f$ to (groups), which then by Theorem 1 may be extended to a functor $A,B \to [A,B]_1$ from $(\mathrm{Ho}\,\mathcal{C})^o \times \mathrm{Ho}\,\mathcal{C}$ to groups, not uniquely but unique up to canonical isomorphism. By the first corollary of Theorem 1 and (4) and (5) the bifunctor $[\cdot,\cdot]_1$ is representable in the first and second variables which proves the theorem. Q.E.D.

<u>Remarks</u>: 1. Σ and Ω are adjoint functors on $\mathrm{Ho}\,\mathcal{C}$ and are unique up to canonical isomorphism. Also for any X, $\Sigma^n X$ $n \geq 1$ is a cogroup object (resp. $\Omega^n X$ is a group object) in $\mathrm{Ho}\,\mathcal{C}$, which is commutative for $n \geq 2$.

2. We shall indulge in the abuse of notation of writing Σ for both the functors on $\mathrm{Ho}\,\mathcal{C}$ of Theorem 2 and writing ΣA for the cofibre of $A \vee A \to A \times I$ when A is in $\mathcal{C}_c$. If we should encounter a situation where this would lead to confusion we shall

denote the former use of Σ by $\underline{\underline{L}}\Sigma$ because it's kind of a left-derived functor in the sense of §4 below. Similarly $\underline{\underline{R}}\Omega$ will be used for the loop functor on $\mathrm{Ho}\,\mathcal{C}$ if necessary.

§3. Fibration and Cofibration Sequences.

In this section we develop another part of the extra structure on $\mathrm{Ho}\,\mathcal{C}$, namely the long exact sequences for fibrations and cofibrations and the Toda bracket operation.

$\mathcal{C}$ denotes a fixed pointed model category in the following.

Let $p:E \to B$ be a fibration where B is fibrant and let $i:F \to E$ be the inclusion of the fibre of p into E. F and E are fibrant by M3. Let $B \xrightarrow{s^B} B^I \xrightarrow{(d_o^B, d_1^B)} B\times B$ be a factorization of Δ_B into a weak equivalence followed by a fibration. We shall construct an object E^I which is nicely related to B^I.

Let $E\times_B B^I$ (resp. $B^I\times_B E$) denote the fibre product of $p:E \to B$ and $d_o^B:B^I \to B$ (resp. $d_1^B:B^I \to B$), and let the fibre product sign $\times_B B^I$ to the left (resp. $B^I\times_B$ to the right) of B^I denote fibre products with d_o^B (resp. d_1^B) in what follows. Let $E \xrightarrow{s^E} E^I \xrightarrow{(d_o^E, p^I, d_1^E)} E\times_B B^I\times_B E$ be a factorization of $(id_E, s^B p, id_E)$ into a weak equivalence followed by a fibration. The notation E^I, s^E, etc. is justified because s^E is a weak equivalence and (d_o^E, d_1^E) is a fibration by M3 since it is the composition of (d_o^E, p^I, d_1^E) and (pr_1, pr_3): $E\times_B B^I\times_B E \to E\times E$, which is the base extension of (d_o^B, d_1^B) by $p\times p$. A similar argument shows that $(d_o^E, p^I):E^I \to E\times_B B^I$ and (p^I, d_1^E) are fibrations.

The map $pr_1:E\times_B B^I \to E$ is the base extension of d_o^B by p and hence is a trivial fibration by M3 and M4. Hence by M5 the fibration $(d_o^E, p^I):E^I \to E\times_B B^I$ is trivial since $id_E =$

$pr_1 \circ (d_o^E, p^I) \circ s_E$. The diagram

$$
(1) \qquad
\begin{array}{ccc}
F \times_E E^I \times_E F & \xhookrightarrow{\;pr_2\;} & E^I \\
\Big\downarrow \pi & & \Big\downarrow (d_o, p^I) \\
F \times \Omega B & \xhookrightarrow{\;1 \times j\;} & E \times_B B^I
\end{array}
$$

is cartesian where $\pi = (pr_1, j^{-1} p^I pr_2)$ and where $j : \Omega B \hookrightarrow B^I$ is as in §2 the fiber of (d_o^B, d_1^B) . Here we are using the followint convention which will be used many times in this section.

<u>Convention</u>: If $\alpha : X \to Y$ is a monomorphism in a category and $\beta : Z \to Y$ is a map, then by $\alpha^{-1}\beta$ we mean the unique map $\gamma : Z \to X$ with $\alpha \circ \gamma = \beta$, if such a map exists.

Returning to the cartesian diagram (1) we have that π is a trivial fibration by M4 and hence in $Ho\mathcal{C}$ (in fact in $Ho\,\mathcal{C}_f$) there is a map

$$
(2) \qquad m : F \times \Omega B \to F
$$

given by the composition $F \times \Omega B \xrightarrow{\gamma(\pi)^{-1}} F \times_E E^I \times_E F \xrightarrow{\gamma(pr_3)} F$.

<u>Proposition 1</u>: The map m is independent of the choice of $p^I : E^I \to B^I$ and is a right action of the group object ΩB on F in $Ho\,\mathcal{C}$.

We first show that m may be defined in another way.

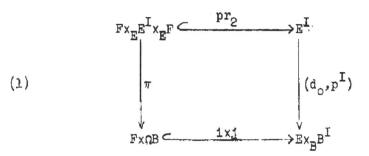

Recall that $[X,Y] = \mathrm{Hom}_{\mathrm{Ho}\,\mathcal{C}}(X,Y)$ and $[X,Y]_1 = [\Sigma X,Y] = [X,\Omega Y]$ where these are the same respectively as $\pi(X,Y)$ and $\pi_1(X,Y)$ if X and cofibrant and Y is fibrant.

Proposition 2: Let A be cofibrant and let the map $m_*:[A,F]\times[A,\Omega B] \to [A,F]$ be denoted by $\alpha,\lambda \to \alpha\cdot\lambda$. If $\alpha \in [A,F]$ is represented by $u:A \to F$, if $\lambda \in [A,\Omega B] = [A,B]_1$ is represented by $h:A\times I \to B$ with $h(\partial_0 + \partial_1) = 0$, and if h' is a dotted arrow in

(3)
$$\begin{array}{ccc} A & \xrightarrow{\ iu\ } & E \\ {\scriptstyle\partial_0}\downarrow & \overset{h'}{\nearrow} & \downarrow{\scriptstyle p} \\ A\times I & \xrightarrow{\ h\ } & B \end{array}$$

then $\alpha\cdot\lambda$ is represented by $i^{-1}h'\partial_1:A \to F$.

Proof: Let $H:A\times I \to B^I$ be a correspondence of h' with $k:A \to B^I$. Let K be a lifting in

$$\begin{array}{ccc} A & \xrightarrow{\ s_E h'\partial_1\ } & E^I \\ {\scriptstyle\partial_1}\downarrow & \overset{K}{\nearrow} & \downarrow{\scriptstyle (d_0^E,\,p^I)} \\ A\times I & \xrightarrow{\ (h',H)\ } & E\times_B B^I \end{array}$$

Picture:

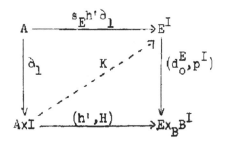

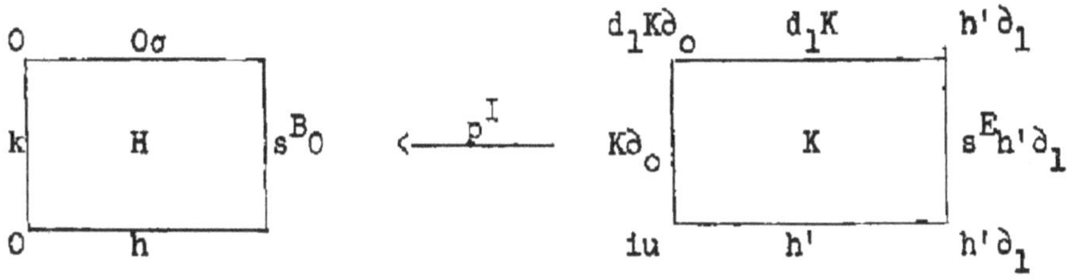

Now $K\partial_o : A \to E^I$ induces a map $K\partial_o : A \to F \times_E E^I \times_E F$ such that $\pi K\partial_o = (u, j^{-1}k)$ (see (1)) and hence by the definition of m we have that $\alpha \cdot \lambda$ is represented by $i^{-1} d_1^E K\partial_o : A \to F$. But $i^{-1} d_1^E K : A \times I \to F$ is a homotopy from $i^{-1} d_1^E K\partial_o$ to $i^{-1} h'\partial_1$ and this proves Proposition 2.

<u>Proof</u> of Prop. 1: Diagram (3) is clearly independent of p^I so m is independent of p^I by Prop. 2. On the other hand, let $\alpha, \lambda, u, h, h'$ be as in Prop. 2, let $\lambda_1 \in [A,B]_1$ be represented by $h_1 : A \times I \to B$ and let h_1' be a dotted arrow in the first diagram

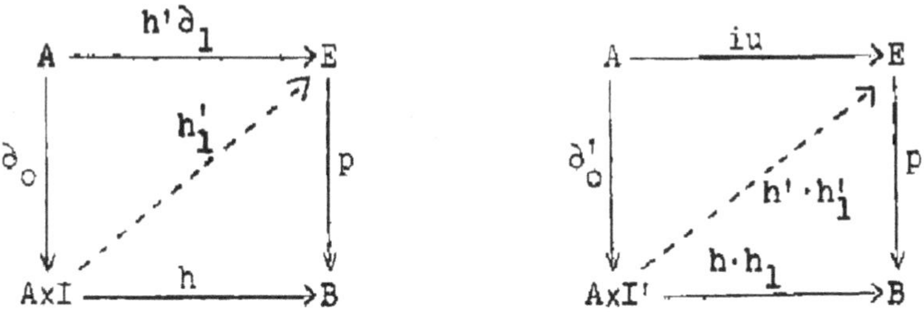

so that $i^{-1} h_1' \partial_1$ represents $(\alpha \cdot \lambda) \cdot \lambda_1$ by Prop. 2. As the composite homotopy $h \cdot h_1$ represents $\lambda \cdot \lambda_1$, the second diagram and Prop. 2 show that $i^{-1}(h' \cdot h_1')\partial_1'$ represents $\alpha \cdot (\lambda \cdot \lambda_1)$. But $(h' \cdot h_1')\partial_1' = h_1' \partial_1$ hence $(\alpha \cdot \lambda) \cdot \lambda_1 = \alpha \cdot (\lambda \cdot \lambda_1)$ and m is an action as claimed. Q.E.D.

<u>Definition 1</u>: By a <u>fibration sequence</u> in $\mathrm{Ho}\mathcal{C}$ we mean a diagram in $\mathrm{Ho}\mathcal{C}$ of the form

$$X \to Y \to Z \qquad X \times \Omega Z \to X$$

which for some fibration $p: E \to B$ in $\mathcal{C}_f$ is isomorphic to the diagram

$$(4) \qquad F \xrightarrow{i} E \xrightarrow{p} B \qquad F \times \Omega B \xrightarrow{m} F$$

constructed above.

<u>Remarks</u>: By dualizing the above construction one may construct a diagram

$$A \to X \to C \qquad C \to C \vee \Sigma A$$

starting from a cofibration u in $\mathcal{C}_c$, where $v: X \to C$ is the cofibre of u and n is a right co-action of the cogroup ΣA on C, and define the notion of a <u>cofibration sequence</u> in $\mathrm{Ho}\mathcal{C}$.

<u>Proposition 3</u>: If (4) is a fibration sequence so is

$$(5) \qquad \Omega B \xrightarrow{\partial} F \xrightarrow{i} E \qquad \Omega B \times \Omega E \xrightarrow{n} \Omega B$$

where ∂ is the composition $\Omega B \xrightarrow{(0,id)} F \times \Omega B \xrightarrow{m} F$ and where $n_*: [A, \Omega B] \times [A, \Omega E] \to [A, \Omega B]$ is given by $(\lambda, \mu) \to ((\Omega p)_* \mu)^{-1} \cdot \lambda$.

<u>Proof</u>: We may assume that (4) is the sequence constructed above from a fibration p. Let $p^I: E^I \to B^I$ be as in the definition of m. Then $pr_1: E \times_B B^I \times_B (*) \to E$ is the base extension

of (d_0^B, d_1^B) by $(p,0):E \to B\times B$ and hence is a fibration; so we get a fibration sequence

(6) $$\Omega B \xrightarrow{(0,j,0)} E\times_B B^I \times_B (*) \xrightarrow{pr_1} E \qquad \Omega B \times \Omega E \xrightarrow{n} \Omega B .$$

We calculate n by Proposition 2; let $\lambda \in [A,\Omega B]$ be represented by $u:A \to \Omega B$, let $\mu \in [A,\Omega E]$ be represented by $h:A\times I \to E$ and let $(h,H,0)$ be a lifting in

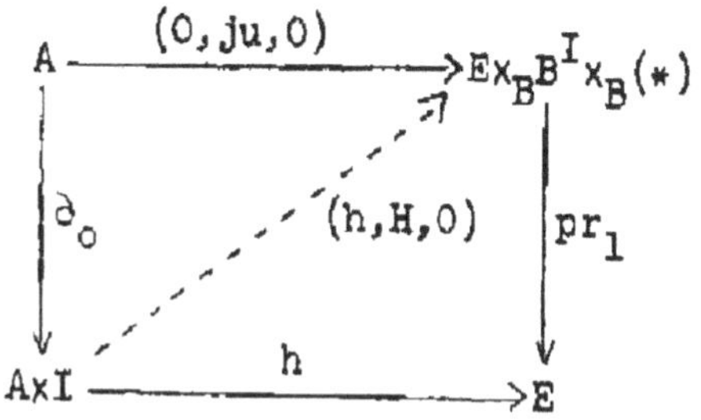

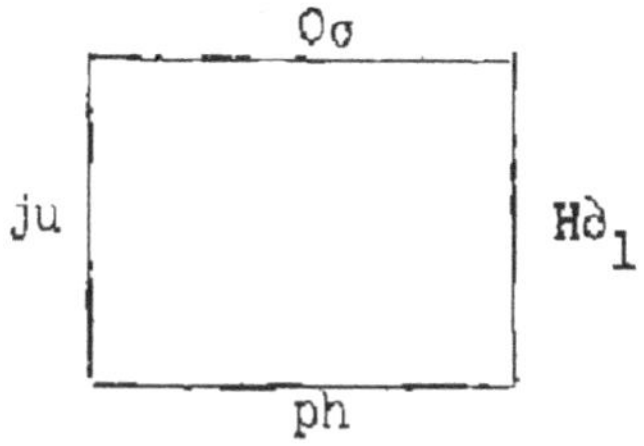

where $H:A\times I \to B^I$ is pictured at the right. By Prop. 2, $j^{-1}H\partial_1$ represents $n_*(\lambda,\mu)$ in $[A,\Omega B]$. Letting $H':A\times I \to B^I$ be a correspondence of $H\partial_1$ with $h':A\times I \to B$, we obtain the correspondence

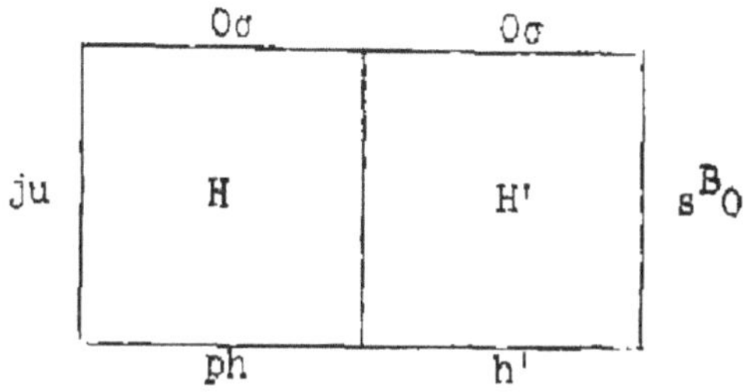

of ju with $ph\cdot h'$, which shows that $\lambda = (\Omega p)_*\mu\cdot n_*(\lambda,\mu)$ or

$n_*(\lambda,\mu) = [(\Omega p)_*\mu]^{-1}\cdot\lambda$. Thus the map n in (6) is the same as that in (5).

The map $f \xrightarrow{(i,0,0)} E\times_B B^I \times_B (*)$ is a weak equivalence by M5 since it may be factored $F \xrightarrow{(s^E i, id)} E^I \times_E F \simeq E^I \times_B (*) \to E\times_B B^I \times_B (*)$ where the second map is a trivial fibration (base extension of $E^I \xrightarrow{(d_0^E, p^I)} E\times_B B^I$) and where the first map is a section of the trivial fibration $E^I \times_E F \xrightarrow{pr_2} F$ (base extension of d_1^E .) We shall show that the diagram in Ho $\mathcal{C}$

(7)

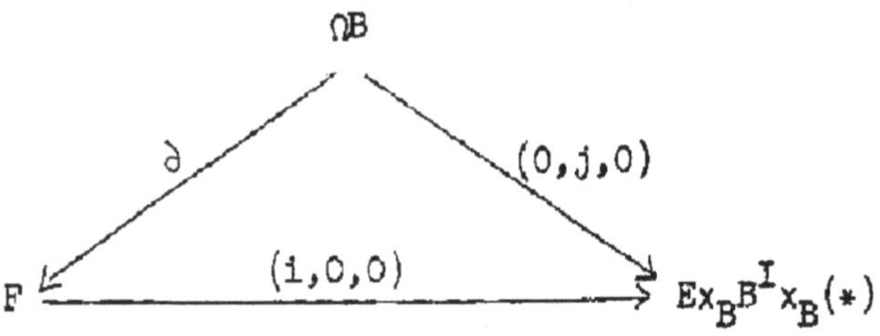

commutes. Let $\lambda \in [A,\Omega B]$ be represented by $k:A \to B^I$ and let $H:A\times I \to B^I$ be a correspondence of k with h . Then $\partial_*\alpha = 0\cdot\alpha$ is represented by $i^{-1}h'\partial_1 : A \to F$ where h' is the dotted arrow in

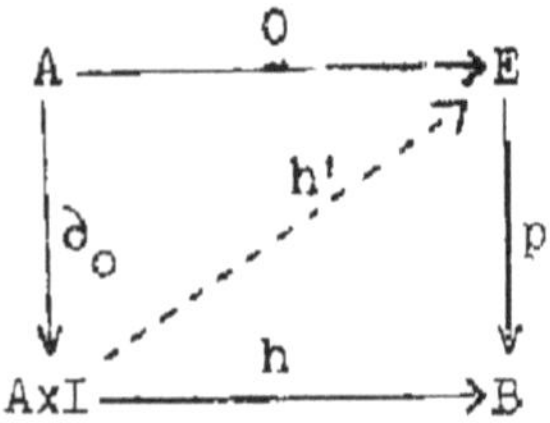

So $(i,0,0)_*\partial_*\lambda$ is represented by $A \xrightarrow{(h'\partial_1,0,0)} E\times_B B^I \times_B (*)$,

$(0,j,0)_*\lambda$ is represented by $A \xrightarrow{(0,k,0)} Ex_B B^I x_B(*)$, and $(h',H,0):A^{\times I} \to Ex_B B^I x_B(*)$ is a left homotopy between these maps, showing that the triangle (7) commutes in $Ho\,\mathcal{C}$. As $pr_1 \circ (i,0,0) = i$ we see that $id_{\Omega B}$, $(i,0,0)$, and id_E give an isomorphism of (5) with the fibration sequence (6), and so by definition (5) is a fibration sequence. Q.E.D.

<u>Proposition 4</u>: Let (4) be a fibration sequence in $Ho\,\mathcal{C}$, let $\partial:\Omega B \to F$ be defined as in Proposition 3 and let A be any object of $Ho\,\mathcal{C}$. Then the sequence

$$\cdots \longrightarrow [A,\Omega^{q+1}B] \xrightarrow{(\Omega^q\partial)_*} [A,\Omega^q F] \xrightarrow{(\Omega^q i)_*} [A,\Omega^q E] \xrightarrow{(\Omega^q p)_*} \cdots$$

$$\cdots \longrightarrow [A,\Omega E] \xrightarrow{(\Omega p)_*} [A,\Omega B] \xrightarrow{\partial_*} [A,F] \xrightarrow{i_*} [A,E] \xrightarrow{p_*} [A,B]$$

is exact in the following sense:

(i) $(p_*)^{-1}\{0\} = \mathrm{Im}\, i_*$

(ii) $i_*\partial_* = 0$ and $i_*\alpha_1 = i_*\alpha_2 \Longleftrightarrow \alpha_2 = \alpha_1 \cdot \lambda$ for some $\lambda \in [A,\Omega B]$

(iii) $\partial_*(\Omega i)_* = 0$ and $\partial_*\lambda_1 = \partial_*\lambda_2 \Longleftrightarrow \lambda_2 = (\Omega p)_*\mu \cdot \lambda_1$ for some $\mu \in [A,\Omega E]$

(iv) The sequence of group homomorphisms from $[A,\Omega E]$ to the left is exact in the usual sense.

The dual proposition for cofibration sequences is

<u>Proposition 4'</u>: Let $A \xrightarrow{u} X \xrightarrow{v} C \xrightarrow{n} C \vee \Sigma A$ be a cofibration

sequence in $\mathrm{Ho}\mathcal{C}$ and let $\partial : C \to \Sigma A$ be $(\mathrm{id}_C + 0) \circ n$. If B is any object in $\mathrm{Ho}\mathcal{C}$, then the sequence

$$\xrightarrow{(\Sigma v)^*} [\Sigma X,B] \xrightarrow{(\Sigma u)^*} [\Sigma A,B] \xrightarrow{\partial^*} [C,B] \xrightarrow{v^*} [X,B] \xrightarrow{u^*} [A,B]$$

is exact in the sense that (i)-(iv) hold with i_*, p_*, ∂_* replaced by v^*, u^*, ∂^* and where the $\cdot$ in (ii) refers to the right action $n^* : [C,B] \times [\Sigma A,B] \to [C,B]$.

Proof of Prop. 4: We may assume (4) is the sequence constructed from the fibration p .

(i) Clearly $pi = 0$. If $p_*\alpha = 0$ represent α by $u : A \to E$, let $h : A \times I \to B$ be such that $h\partial_0 = pu$, $h\partial_1 = 0$. By the covering homotopy theorem (dual of Cor. of Lemma 2, §1 we may cover h by $k : A \times I \to E$ with $\partial_0 k = u$. Then if β is represented by $i^{-1}k\partial_1$ we have $i_*\beta = \alpha$.

(ii) With the notation of Prop. 2, we have that h' is a homotopy from iu which represents $i_*\alpha$ to $h'\partial_1$ which represents $i_*(\alpha \cdot \lambda)$. Hence $i_*(\alpha \cdot \lambda) = i_*\alpha$ and in particular $i_*\partial_*\lambda = i_*(0 \cdot \lambda) = i_*0 = 0$, so $i_*\partial_* = 0$. Conversely given α_1, α_2 with $i_*\alpha_1 = i_*\alpha_2$, represent α_i by u_i , $i = 1,2$, let $h : A \times I \to E$ be such that $h\partial_0 = iu_1$, $h\partial_1 = iu_2$ whence if λ is the class of ph , $\alpha_1 \cdot \lambda = \alpha_2$ by Prop. 2.

(iii) follows from (ii) and Prop. 3, and (iv) by repeated use of Proposition 3. Q.E.D.

Proposition 5: The class of fibration sequences in $\mathrm{Ho}\mathcal{C}$ has the following properties:

(i) Any map $f: X \to Y$ may be embedded in a fibration sequence $F \to X \xrightarrow{f} Y$, $F \times \Omega Y \to F$.

(ii) Given a diagram of solid arrows

$$(8) \qquad \begin{array}{ccccc} F & \xrightarrow{i} & E & \xrightarrow{p} & B \\ \Big\downarrow{\scriptstyle \gamma} & & \Big\downarrow{\scriptstyle \beta} & & \Big\downarrow{\scriptstyle \alpha} \\ F' & \xrightarrow{i'} & E' & \xrightarrow{p'} & B' \end{array} \qquad\qquad \begin{array}{ccc} F \times \Omega B & \xrightarrow{m} & F \\ \Big\downarrow{\scriptstyle \gamma \times \Omega\alpha} & & \Big\downarrow{\scriptstyle \gamma} \\ F' \times \Omega B' & \xrightarrow{m'} & F' \end{array}$$

where the rows are fibration sequences, the dotted arrow γ exists.

(iii) In any diagram (8) where the rows are fibration sequences, if α and β are isomorphisms so is γ.

(iv) Proposition 3.

Remark: Proposition 4 gives the analogues for fibration sequences of all the non-trivial axioms for the triangles in a triangulated category (see Verdier [11] or Hartshorne [9]) except the octahedral axiom. The analogue of that axiom holds also, but as far as the author knows, it's not worth the trouble required to write it down.

Proof: (i) Any map in $\mathrm{Ho}\,\mathcal{C}$ is isomorphic to a fibration of objects in $\mathcal{C}_{cf}$.

(iii) If A is any object in $\mathrm{Ho}\,\mathcal{C}$, then Prop. 4 gives a diagram

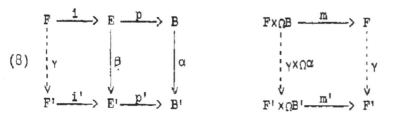

$$\begin{array}{ccccccccc}
[A,\Omega E] & \longrightarrow & [A,\Omega B] & \longrightarrow & [A,F] & \longrightarrow & [A,E] & \longrightarrow & [A,B] \\
\downarrow \wr & & \downarrow \wr & & \downarrow \gamma_* & & \downarrow \wr & & \downarrow \wr \\
[A,\Omega E'] & \longrightarrow & [A,\Omega B'] & \longrightarrow & [A,F'] & \longrightarrow & [A,E'] & \longrightarrow & [A,B']
\end{array}$$

where the rows are "exact" in the sense that (i)-(iii) of Prop. 4

hold. However this is enough to conclude by the usual 5 lemma argument that $\gamma_*:[A,F] \to [A,F']$ is a bijection for all A and hence that γ is an isomorphism.

(ii) We may suppose by replacing the diagram (8) by an isomorphic diagram if necessary that the rows are constructed in the standard way from fibrations p and p' in $\mathcal{C}_f$. Let $\tilde{B} \xrightarrow{u} B$ be a trivial fibration with $\tilde{B}$ cofibrant and let $\tilde{E} \xrightarrow{v} E\times_B\tilde{B}$ be a trivial fibration with $\tilde{E}$ cofibrant. By M4 $pr_1:E\times_B\tilde{B} \to E$ is a trivial fibration and $pr_2:E\times_B\tilde{B} \to \tilde{B}$ is a fibration so we obtain a diagram

$$\begin{array}{ccccc}
\tilde{F} & \xrightarrow{\tilde{i}} & \tilde{E} & \xrightarrow{pr_2 v} & \tilde{B} \\
\downarrow{\scriptstyle \varepsilon} & & \downarrow{\scriptstyle pr_1 v} & & \downarrow{\scriptstyle u} \\
F & \xrightarrow{i} & E & \xrightarrow{p} & B
\end{array}$$

in $\mathcal{C}$, where pr_1v and u are weak equivalences. It follows easily from the calculation given in Prop. 2, that

$$\begin{array}{ccc}
\tilde{F}\times\Omega\tilde{B} & \xrightarrow{\tilde{m}} & \tilde{F} \\
\downarrow & & \downarrow \\
F\times\Omega B & \xrightarrow{m} & F
\end{array}$$

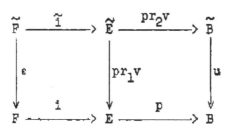

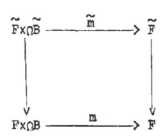

commutes. Hence by (iii) the $\sim$ sequence is isomorphic to first row of (8) and so we may suppose that the rows of (8) are not only constructed in the standard way from fibrations p and p' but that E and B are in $\mathcal{C}_{cf}$. Then by Theorem 1 α and β are represented by maps u and v in $\mathcal{C}$ with $p'v \sim up$. As E is cofibrant, we may by the corollary of Lemma 2, §1, modify v, so that $p'v = up$. Then we may take $\gamma: F \to F'$ in (8) to be the map in $\mathcal{C}$ induced by v. The first part of (8) commutes clearly, and the second square may be shown to commute in $\mathrm{Ho}\,\mathcal{C}$ by use of Proposition 2. This proves (ii). Q.E.D.

The dual proposition for cofibration sequences is left to the reader.

The following proposition will be used in the definition of the Toda bracket

<u>Proposition 6</u>: Let

(9) 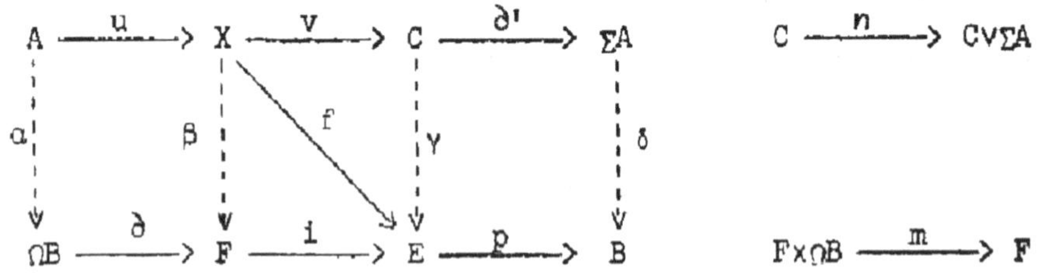

be a solid arrow diagram in $\mathrm{Ho}\,\mathcal{C}$ where the first row except for ∂' is a cofibration sequence, and where the second row except for ∂ is a fibration sequence. We suppose that $\partial' = (\mathrm{id}_C + 0)\circ n$ and $\partial = m\circ(0, \mathrm{id}_{\Omega B})$ as in Propositions 4 and 4'. Suppose that

$fu = 0$ and $pf = 0$. Then dotted arrow $\alpha,\beta,\gamma,\delta$ exist and the set of possibilities for α forms a left $\Omega p_*[A,\Omega E]$ -right $u^*[X,\Omega B]$ double coset in $[A,\Omega B]$ and the set of possibilities for δ forms a left $(\Sigma u)^*[\Sigma X,B]$ -right $p_*[\Sigma A,E]$ double coset in $[\Sigma A,B]$. Furthermore under the identification $[A,\Omega B] = [\Sigma A,B]$ the first double coset is the inverse of the second.

Proof: By Prop. 4 $pf = 0 \Longrightarrow \exists\beta: X \to F$ with $f = i\beta$. Similarly $i\beta u = 0 \Longrightarrow \exists\alpha$ with $\partial\alpha = \beta u$. Hence α,β exist. Suppose that α',β' are other maps. By the exact sequence of Prop. 4 $\beta' = \beta\cdot\lambda$ for some $\lambda \in [X,\Omega B]$. More precisely $\beta' = m_\circ(\beta,\lambda)$ hence $\partial\alpha' = \beta' u = m_\circ(\beta,\lambda)u = m(\beta u,\lambda u) = m(\partial\alpha,\lambda u) = \partial\alpha\cdot(\lambda u) = (0\cdot\alpha)\cdot\lambda u = 0(\alpha\cdot\lambda u) = \partial(\alpha\cdot\lambda u)$. By exactness $\alpha' = (\Omega p)_*\mu\cdot\alpha\cdot\lambda u = (\Omega p)_*\mu\cdot\alpha\cdot u^*(\lambda)$ and so α' lies in the double coset $\Omega p_*[A,\Omega E]\cdot\alpha\cdot u^*[X,\Omega B]$. As μ and λ may be arbitrary we see that any element of this double coset may be an α'. Dual assertions hold for γ and δ and so the first statement of the proposition is proved.

To prove the second assertion we must construct $\alpha,\beta,\gamma,\delta$ so that α corresponds to δ^{-1}. We may assume that u is a cofibration of cofibrant objects, that p is a fibration of fibrant objects and that the top and bottom rows of (1) are constructed as above. In this case Theorem 1 shows that the map f in Ho$\mathcal{C}$ may be represented by a map in $\mathcal{C}$ which we shall denote again by f. Now $pf \sim 0$ and as X is cofibrant and u is a fibration we may by the corollary to Lemma 2, §1 lift this homotopy to E and so assume that $pf = 0$. (We may not, however, simultaneously assume that $fu = 0$.) Let $h: A \times I \to E$ be such that $h\partial_o = fu$, $h\partial_1 = 0$ and consider the following diagram

(10)

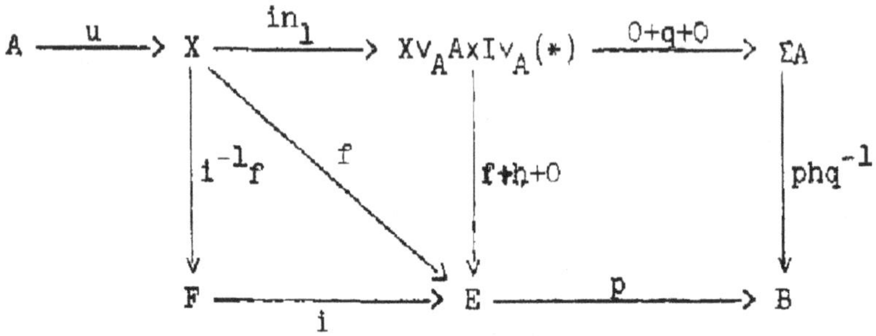

where $q: A \times I \to \Sigma A$ is the cokernel of $A \vee A \to A \times I$ and where we extend to epimorphisms the convention for monomorphisms introduced at the beginning of this section so that phq^{-1} is the unique map such that $(phq^{-1})\, q = ph$. Now the top line of (10) is isomorphic in Ho$\mathcal{E}$ to the top line of the first part of (9)--see the proof of Proposition 3 especially the homotopy commutativity of (7) for the dual considerations. Consequently by means of this isomorphism we may define β in (9) to be represented by $i^{-1}f$, γ by $f + h + 0$, and δ by phq^{-1} . But we also have the diagram

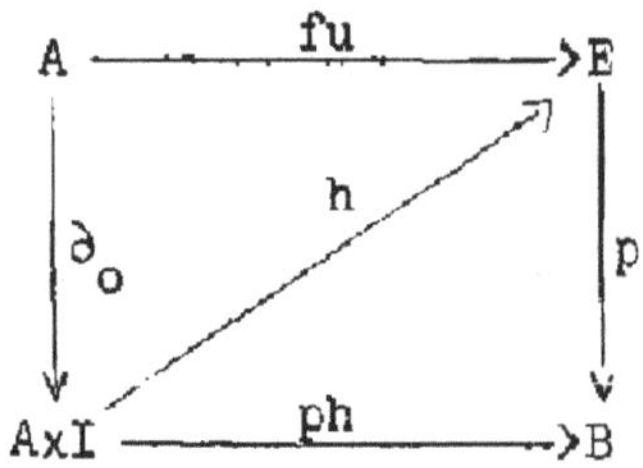

$h\partial_1 = 0$

which by Prop. 2, §3 shows that $\beta u \cdot \delta = 0$ since $i^{-1}fu$ represents βu in (1). Hence $\beta u = 0 \cdot \delta^{-1} = \partial(\delta^{-1})$ and we may take α in (1) to be δ^{-1}.

<u>Definition 2</u>: Let $A \xrightarrow{u} X \xrightarrow{f} E \xrightarrow{p} B$ be three maps in $\mathrm{Ho}\mathcal{C}$ such that $fu = pf = 0$. Form a solid arrow diagram

(11)

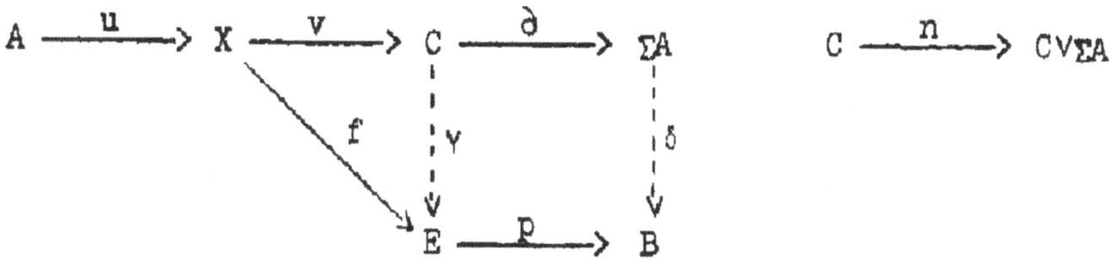

by choosing by Prop. 5(i) for the first row a cofibration sequence containing u, and then fill in the dotted arrows as in Prop. 6. The set of possibilities for δ is as in Prop. 6 a left $(\Sigma u)^*[\Sigma X,B]$ -right $p_*[\Sigma A,E]$ double coset in $[\Sigma A,B]$ which is called the <u>Toda bracket</u> of $u,f,$ and p, and is denoted $\langle u,f,p\rangle$.

<u>Remarks</u>: 1. The Toda bracket is independent of the choice of the top row of (3) by Prop. 5(ii) and (iii).

2. The Toda bracket $\langle u,f,p\rangle$ may also be computed by choosing a solid arrow diagram

(12)

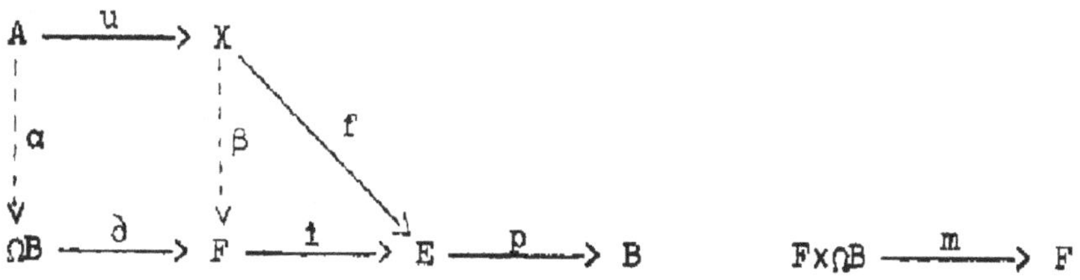

where the bottom row comes from a fibration sequence, and filling in the dotted arrows. By Proposition 6 we have $\langle u,f,p\rangle =$ $(\Sigma u)^*[\Sigma Y,B]\cdot\alpha^{-1}\ .\ p_*[\Sigma A,B] \subseteq [\Sigma A,B]$.

4. Equivalence of Homotopy Theories

We begin with some general categorical considerations.

Definition 1: Let $\gamma: \underline{A} \longrightarrow \underline{A}'$ and $F: \underline{A} \longrightarrow \underline{B}$ be two functors. By the left-derived functor of F with respect to γ we mean a functor $L^{\gamma}F: \underline{A}' \longrightarrow \underline{B}$ with a natural transformation $\varepsilon: L^{\gamma}F \circ \gamma \longrightarrow F$ having the following universal property: Given any $G: \underline{A}' \longrightarrow B$ and natural transformation $\zeta: G \circ \gamma \longrightarrow F$ there is a unique natural transformation $\theta: G \longrightarrow L^{\gamma}F$ such that

(1)

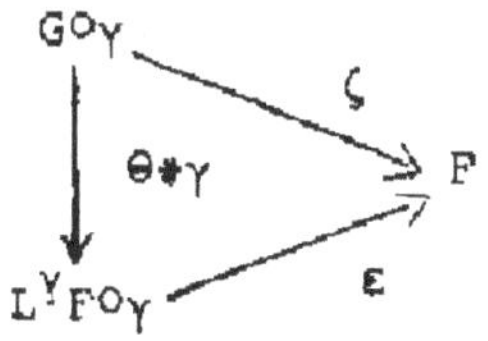

commutes.

Remarks: 1. $L^{\gamma}F$ is the functor from $\underline{A}'$ to $\underline{B}$ such that $L^{\gamma}F \circ \gamma$ is closest to F from the left. Similarly we may define the right-derived functor of F with respect to γ to be "the" functor $R^{\gamma}F: \underline{A}' \longrightarrow \underline{B}$ with a natural transformation $\eta: F \longrightarrow R^{\gamma}F \circ \gamma$ which is closest to F from the right.

2. The terminology left-derived functor comes from Verdier's treatment of homological algebra[11]. In that case $\underline{A}$ is the category K(A), where A is an abelian category, γ is the localization $K(A) \longrightarrow D(A)$, $F: K(A) \longrightarrow \underline{B}$ is a cohomological functor from K(A) to an abelian category $\underline{B}$, and $L^{\gamma}F$, $R^{\gamma}F$ are what Verdier calls the left and right

derived functors of F.

3. We shall be concerned only with the case where $\underline{A}$ is a model category $\underline{C}$ and γ is the localization functor $\gamma: \underline{C} \longrightarrow Ho\underline{C}$. In this case we will write just LF.

4. If $\underline{C}$ is a model category and $F: \underline{C} \longrightarrow \underline{B}$ is a functor then it is clear that $\varepsilon: LF\circ\gamma \longrightarrow F$ is an isomorphism if and only if F carries weak equivalences in $\underline{C}$ into isomorphisms in $\underline{B}$. In this case we may assume that LF is induced by F in the sense that LF is the unique functor $Ho\underline{C} \longrightarrow \underline{B}$ with $LF\circ\gamma = F$. Moreover RF=LF.

Proposition 1: Let $F: \underline{C} \longrightarrow \underline{B}$ be a functor where $\underline{C}$ is a model category. Suppose that F carries weak equivalences in $\underline{C}_c$ into isomorphisms in $\underline{B}$. Then $LF: Ho\underline{C} \longrightarrow \underline{B}$ exists. Furthermore $\varepsilon(X): LF(X) \longrightarrow F(X)$ is an isomorphism if X is cofibrant.

Proof: Let $X \longrightarrow Q(X)$, $f \longrightarrow Q(f)$, $p_X: Q(X) \longrightarrow X$ be as in the proof of theorem 1, §1, so that Q induces a well-defined functor $\overline{Q}: \underline{C} \longrightarrow \pi\underline{C}_c$. By lemma 8(ii), §1, $X \longrightarrow FQX, f \longrightarrow FQ(f)$ is a functor $FQ: \underline{C} \longrightarrow \underline{B}$ which induces a functor $LF: Ho\underline{C} \longrightarrow \underline{B}$, since Q(f) is a weak equivalence if f is. Let $\varepsilon: LF\circ\gamma \longrightarrow F$ be the natural transformation given by $\varepsilon(X) = F(p_X): FQX \longrightarrow FX$. To show that ε has the universal property of definition 1, let $\zeta: G\circ\gamma \longrightarrow F$ where $G: Ho\underline{C} \longrightarrow \underline{B}$. Define $\theta(X): G(X) \longrightarrow LF(X)$ to be the composition

$$G(X) \xrightarrow{\;\;G(\gamma(p_X)^{-1})\;\;} GQX \xrightarrow{\;\;\zeta\;\;} FQX = LF(X).$$ It is clear

that Θ is a natural transformation $G \circ \gamma \longrightarrow LF \circ \gamma$, and

since every map is $Ho\underline{C}$ is a finite composition of maps

$\gamma(f)$ or $\gamma(s)^{-1}$, Θ is a natural transformation $\Theta: G \longrightarrow LF$.

The diagram

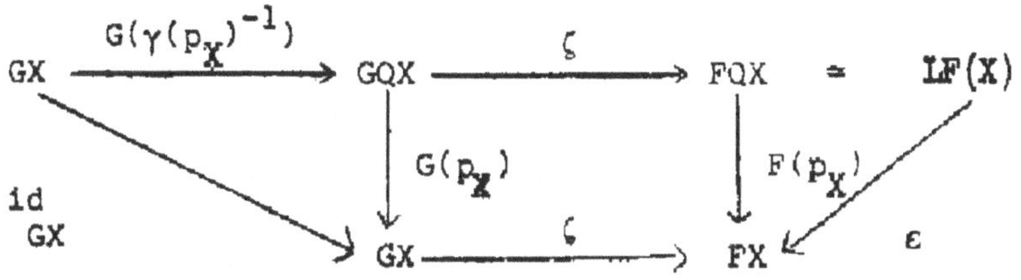

shows that $\varepsilon(\Theta * \gamma) = \zeta$. The uniqueness of $\Theta: G \longrightarrow LF$

is clear since it is determined by   on $Ho\underline{C}_c = Ho\underline{C}$ and

so ε has the required universal property. Finally if X

is cofibrant $LFX=FQX=FX$ and $\varepsilon(X) = id_{F(X)}$.   QED.

<u>Definition 2</u>: Let $F: \underline{C} \longrightarrow \underline{C}'$ be a functor where

$\underline{C}$ and $\underline{C}'$ are model categories. By the <u>total left-derived</u>

<u>functor</u> of F we mean the functor $\underline{L}F: Ho\underline{C} \longrightarrow Ho\underline{C}'$ given

by $\underline{L}F = L^{\gamma}(\gamma' \circ F)$ where $\gamma: \underline{C} \longrightarrow Ho\underline{C}$ and $\gamma': \underline{C}' \longrightarrow Ho\underline{C}'$

are the localization functors.

<u>Remark</u>: The diagram

(2)
$$\begin{array}{ccc} \underline{C} & \xrightarrow{\;F\;} & \underline{C}' \\ \gamma \downarrow & & \downarrow \gamma' \\ Ho\underline{C} & \xrightarrow{\;\underline{L}F\;} & Ho\underline{C} \end{array}$$

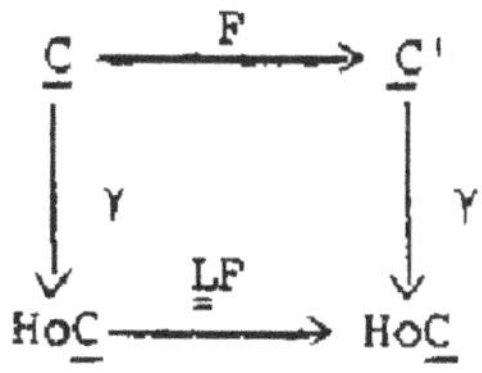

does <u>not</u> commute, but rather there is a natural transformation ε: $\underline{\underline{L}}F \circ \gamma \longrightarrow \gamma' \circ F$ such that the pair $(\underline{\underline{L}}F, \varepsilon)$ comes as close to making (1) commutative as possible.

<u>Corollary</u>: If F carries weak equivalence in $\underline{C}_c$ into weak equivalences $\underline{C}'$, then $\underline{\underline{L}}F$: Ho$\underline{C} \longrightarrow$ Ho$\underline{C}'$ exists and $\varepsilon(X)$: $\underline{\underline{L}}F(X) \longrightarrow F(X)$ is an isomorphism in Ho$\underline{C}'$ for X cofibrant.

<u>Proposition 2</u>: Let $\underline{C}$ and $\underline{C}$ be pointed model categories with suspension functors Σ and Σ' on Ho$\underline{C}$ and Ho$\underline{C}'$, respectively. Let F: $\underline{C} \longrightarrow \underline{C}'$ be a functor which is right exact (i.e. compatible with finite inductive limits), which carries cofibrations in $\underline{C}$ into cofibrations in $\underline{C}'$, and which carries weak equivalences in $\underline{C}_c$ into weak equivalences in $\underline{C}'$. Then $\underline{\underline{L}}F$ is compatible with finite direct sums, there is a canonical isomorphism of functors $\underline{\underline{L}}F \circ \Sigma \simeq \Sigma' \circ \underline{\underline{L}}F$, and with respect to this isomorphism $\underline{\underline{L}}F$ carries cofibration sequences in Ho$\underline{C}$ into cofibration sequences in Ho$\underline{C}'$.

<u>Proof</u>: $\underline{\underline{L}}F$ exists by proposition 1 and we may assume that $\underline{\underline{L}}F(A) = F(A)$ if A is cofibrant. If A_1 and A_2 are in $\underline{C}_c$ then $A_1 \vee A_2$, the direct sum of A_1 and A_2 in $\underline{C}$, is also the direct sum of A_1 and A_2 in Ho$\underline{C}$. By assumption $F(\underline{C}_c) \subset \underline{C}'_c$ and so $\underline{\underline{L}}F(A_1 \vee A_2) = F(A_1 \vee A_2) = F(A_1) \vee F(A_2) = \underline{\underline{L}}F(A_1) \vee \underline{\underline{L}}F(A_2)$ where the last v means direct sum in Ho$\underline{C}'$.

This proves the first assertion about F. Next observe that if A is cofibrant, then for a given object AxI we have that $F(A)vF(A) \xrightarrow{F(\partial_0)+F(\partial_1)} F(AxI) \xrightarrow{F(\sigma)} F(A)$ is a factorization of $\nabla_{F(A)}$ into the cofibration $F(\partial_0)+F(\partial_1) = F(\partial_0+\partial_1)$ followed by the weak equivalence $F(\sigma)$. Hence $F(AxI) = FA \times I$ and since F is compatible with cofibre products $F(\Sigma A) = \Sigma F(A)$. As F(A) is cofibrant $\Sigma(F(A))$ represents $\Sigma(F(A))$ in Ho$\underline{C}$ and so the second assertion is proved. Finally note that if $i: A \rightarrowtail B$ is a cofibration in $\underline{C}_c$ and $AxI \xrightarrow{ixI} BxI$ is a compatible choice in the dual sense that $p^I: E^I \to B^I$ was a compatible choice in §3, then $F(AxI) \to F(BxI)$ is also a compatible choice for $FAxI \to FBxI$. It follows that F carries the diagram in $\underline{C}_c$

$$A \xrightarrow{i} B \xrightarrow{q} C \qquad C \xrightarrow{in_1} C \underset{B}{v} BxI \underset{B}{v} C \xleftarrow{\xi} Cv A$$

where ξ is a weak equivalence into a similar diagram with A replaced by FA, etc. This proves the last assertion about $\underline{\underline{L}}F$. QED.

<u>Theorem 3</u>: Let $\underline{C}$ and $\underline{C}'$ be model categories and let

$$\underline{C} \underset{R}{\overset{L}{\rightleftarrows}} \underline{C}'$$

be a pair of adjoint functors, L being the left and R the right adjoint functor. Suppose that L preserves cofibrations and that L carries weak equivalences in $\underline{C}_c$ into weak equivalences in $\underline{C}'$. Also suppose that R preserves fibrations and that R carries weak equivalences in $\underline{C}'_f$ into weak equivalences in $\underline{C}$. Then the functors

$$\mathrm{Ho}\underline{C} \underset{\underline{\underline{R}}(R)}{\overset{\underline{\underline{L}}(L)}{\rightleftarrows}} \mathrm{Ho}\underline{C}'$$

are canonically adjoint. Suppose in addition for X in $\underline{C}_c$ and Y in $\underline{C}_f$ that a map $LX \longrightarrow Y$ is a weak equivalence if and only if the associated map $X \longrightarrow RY$ is a weak equivalence. Then the adjunction morphisms $\mathrm{id} \longrightarrow \underline{\underline{L}}(L) \circ \underline{\underline{R}}(R)$ and $\underline{\underline{R}}(R) \circ \underline{\underline{L}}(L) \longrightarrow \mathrm{id}$ are isomorphisms so the categories $\mathrm{Ho}\underline{C}$ and $\mathrm{Ho}\underline{C}'$ are equivalent. Furthermore if $\underline{C}$ and $\underline{C}'$ are pointed then these equivalences $\underline{\underline{L}}(L)$ and $\underline{\underline{R}}(R)$ are compatible with the suspension and loop functors and the fibration and cofibration sequences in $\mathrm{Ho}\underline{C}$ and $\mathrm{Ho}\underline{C}'$.

Proof: For simplicity we write $\underline{\underline{L}}$ instead of $\underline{\underline{L}}(L)$ and we use Grothendieck's notation $u^\flat: X \longrightarrow RY$ (resp $v^\#: LX \longrightarrow Y$) to denote the map corresponding to $u: LX \longrightarrow Y$ (resp $v: X \longrightarrow RY$). If X is in $\underline{C}_c$ and Y is in $\underline{C}_f$, then we saw in the proof of proposition 1 that $L(X \times I) = LX \times I$.

Hence to any left homotopy $h: X\times I \longrightarrow RY$ between f and g there corresponds the homotopy $H^b: LX\times I \longrightarrow Y$ between f^b and g^b and so $[X,RY] = [LX,Y]$. Hence if $X \mapsto Q(X)$ etc. is as in the proof of theorem 1, §1 and $Y \mapsto R'(Y)$, $f \mapsto R'(f)$, $i_Y: Y \longrightarrow R'(Y)$ is the functor-up-to-homotopy of theorem 1 for the category $\underline{C}'$ we have the isomorphisms

$$(3)\quad \mathrm{Hom}_{\mathrm{Ho}\underline{C}'}(\underline{\underline{L}}X,Y) \simeq [LQX,R'Y] \simeq [QX,RR'Y] \simeq \mathrm{Hom}_{\mathrm{Ho}\underline{C}}(X,\underline{\underline{R}}Y)$$

where the first and last isomorphisms come from the construction of $\underline{\underline{L}}$ and $\underline{\underline{R}}$ given above in proposition 1. The isomorphisms (3) are clearly functorial as (X,Y) runs over $\underline{C}^0 \times \underline{C}^1$, and hence as every map in $\mathrm{Ho}\underline{C}$ is a finite composition of maps of the form $\gamma(f)$ or $\gamma(s)^{-1}$, (3) is functorial as (X,Y) runs over $(\mathrm{Ho}\underline{C})^{\circ} \times (\mathrm{Ho}\underline{C}')$ proving that $\underline{\underline{L}}$ and $\underline{\underline{R}}$ are adjoint.

Suppose now that for X in $\underline{C}_c$ and Y in C'_f, $f: X \longrightarrow RY$ is a weak equivalence iff $f^{\#}: LX \longrightarrow Y$ is a weak equivalence so $X \xrightarrow{(i_{LX})^b} RR'(LX)$ is a weak equivalence. But by proposition 1, $RR'LX = \underline{\underline{R}}\,\underline{\underline{L}}X$ and by examining (3) we see that $\gamma((i_{LX})^b): X \longrightarrow RR'(LX)$ is the adjunction map $X \longrightarrow \underline{\underline{R}}\,\underline{\underline{L}}X$. Hence $X \xrightarrow{\sim} \underline{\underline{R}}\,\underline{\underline{L}}X$ for all X in $\mathrm{Ho}\underline{C}_c$ and hence in $\mathrm{Ho}\underline{C}$. Similarly $\underline{\underline{L}}\,\underline{\underline{R}} \xrightarrow{\sim} \mathrm{id}$ which proves the second assertion of the theorem.

If $\underline{C}$ and $\underline{C}'$ are pointed we have by proposition 2 and its dual that $\underline{\underline{L}}\Sigma \simeq \Sigma'\underline{\underline{L}}$ and $\Omega\underline{\underline{R}} \simeq \underline{\underline{R}}\Omega'$. Hence $\underline{\underline{R}}\Sigma' \simeq \underline{\underline{R}}\Sigma'\underline{\underline{L}}\underline{\underline{R}} \simeq \underline{\underline{R}}\underline{\underline{L}}\Sigma\underline{\underline{R}} \simeq \Sigma\underline{\underline{R}}$ and similarly $\underline{\underline{L}}$ preserves loop functors. Also by proposition 2 $\underline{\underline{L}}$ preserves cofibration sequences and $\underline{\underline{R}}$ preserves fibration sequences. Suppose that $\varepsilon = \{F \xrightarrow{i} E \xrightarrow{p} B,\ \Omega B\times F \xrightarrow{n} F\}$ is a fibration sequence in $\mathrm{Ho}\underline{C}$. Then we may embed the map $\underline{\underline{L}}E \longrightarrow \underline{\underline{L}}B$ in a fibration sequence ε' of $\mathrm{Ho}\underline{C}'$ by proposition 5 (i), §3 and the image $\underline{\underline{R}}\varepsilon'$ of the sequence under $\underline{\underline{R}}$ is a fibration sequence which is isomorphic to ε by proposition 5,(ii) and (iii), §3. Hence $\varepsilon' \simeq \underline{\underline{L}}\varepsilon$ and $\underline{\underline{L}}$ preserves fibration sequences. Similarly $\underline{\underline{R}}$ preserves cofibration sequences. QED.

Examples 1: Let $\underline{A}$ be an abelian category with enough projectives and injectives and let $\underline{C}$ and $\underline{C}'$ be the two model categories which have $C_b(\underline{A})$ as underlying category described in Remark 3 following theorem 1, §1. Then the identity functor gives a pair of adjoint functor $\underline{C} \rightleftarrows \underline{C}'$ satisfying the conditions of the theorem. The theorem implies that cofibration and fibration sequences constructed from both categories coincide which is clear since they coincide with Verdier's triangles.

2. Let $\underline{C}'$ = (spaces) $\underline{C}$ = (ss sets) as in examples A and C, §1, and let L be the geometric realization functor, and R the singular complex functor. Then theorem 3 applies because of [Milnor [15]] and so the cofibration

sequences in the homotopy categories of ss sets of spaces coincide. This is not entirely trivial since the singular functor does not commute with the operation of taking the cofiber of a map.

<u>Remark</u>: We recall our vague definition of the homotopy theory associated to a model category, namely the category Ho$\underline{C}$ with all extra structure which comes by performing constructions in $\underline{C}$. In §2 and §3 we gave the most important examples of that extra structure and Theorem 3 gives a criterion which shows when the homotopy theories coming from different model categories coincide, at least when only the structure of §2 and §3 is concerned. There are other kinds of structure, e.g. higher order operations ([18] [17]), which are not included in theorem 3, and it seems reasonable to conjecture that this extra structure is preserved under the conditions of theorem 3.

5. Closed Model Categories

We will say that a map i: A --> B has the *left lifting property* with respect to a class S of maps in a category $\underline{C}$ if the dotted arrow exists in any diagram of the form

(1)

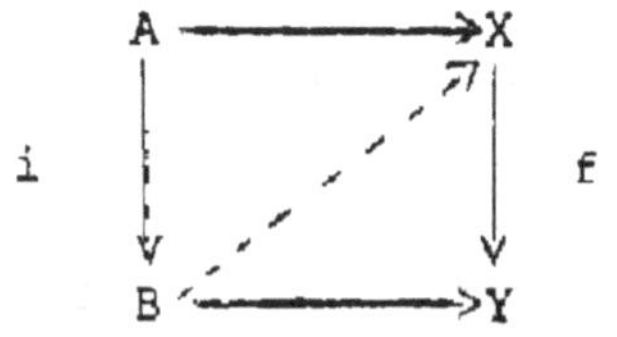

where f is in the class S. Similarly f has the *right lifting property* with respect to S if the dotted arrow exists in any diagram of the form (1) where i is in S.

Definition 1: A model category $\underline{C}$ is said to be *closed* if it satisfies the axiom

M6: Any two of the following classes of maps in $\underline{C}$ - the fibrations, cofibrations, and weak equivalences - determine the third by the following rules:

a) A map is a fibration <==> it has the right lifting property with respect to the maps which are both cofibrations and weak equivalences

b) A map is a cofibration <==> it has the left lifting property with respect to the maps which are both fibrations and weak equivalences.

c) A map f is a weak equivalence <==> f = uv where

v has the left lifting property with respect to the class of fibrations and u has the right lifting property with respect to the class of cofibrations.

Remarks: 1. It is clear that M6 implies M1, M3, and M4. Hence a closed model category may be defined using axioms M0, M2, M5, and M6.

2. Examples A, B, and C of §1 are all closed model categories (see proposition 2 below). Model categories which are not closed may be constructed by reducing the class of cofibrations but keeping M2, M3, and M4 valid. For example, take example B, §1, where $\underline{A}$ is the category of left R modules, R a ring, and define cofibrations to be injective maps f in $C_+(\underline{A})$ such that Coker f is a complex of free R modules.

In the following $\underline{C}$ is a fixed model category and we retain the notations of the previous sections.

Lemma 1: Let $p: X \to Y$ be a fibration in $\underline{C}_{cf}$. The following are equivalent.

(i) p has the right lifting property with respect to the cofibrations.

(ii) p is the dual of a strong deformation retract map in the following precise sense: there is a map $t: Y \to X$ with $pt = id_Y$ and there is a homotopy $h: X \times I \to X$ from tp to id_X with $ph = p\sigma$.

(iii) $\gamma(p)$ is an isomorphism.

<u>Proof</u>: (i) ==> (ii) One lifts successively in

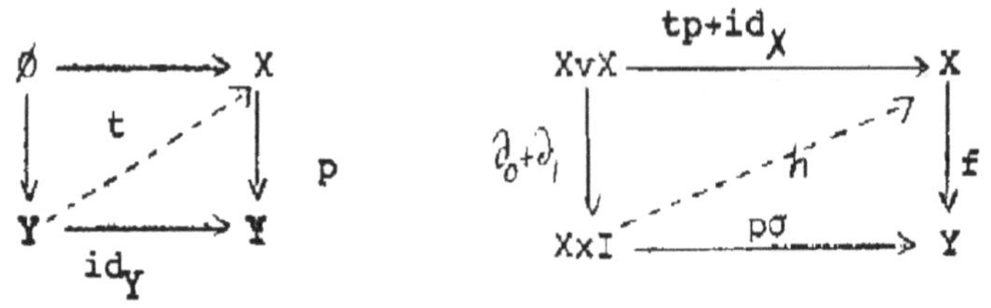

(ii) ==> (i) Let p^I: X^I --> Y^I be a compatible choice of path objects for X and Y as in the beginning of §2 and let Q be a lifting in

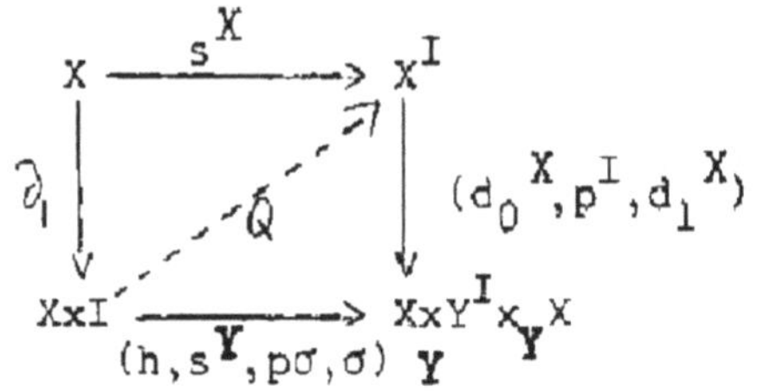

Then $k = Q\partial_0$: X --> X^I is a right homotopy from tp to id_X with $p^I k = s^Y p$. Given the first diagram

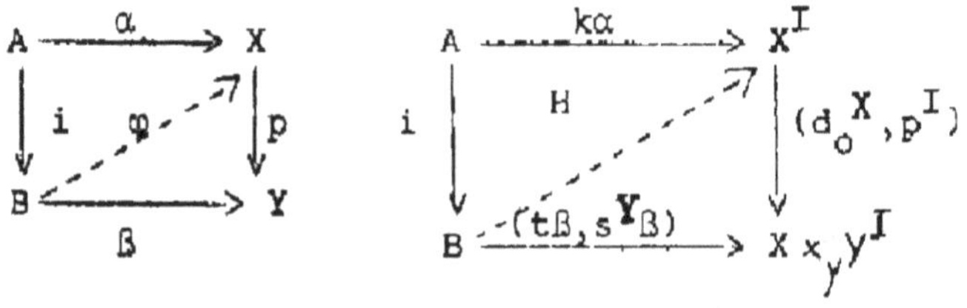

the dotted arrow φ may be constructed by choosing a dotted arrow H in the second and setting $\varphi = d_1{}^X H$.

(ii) ==> (iii). t is a homotopy inverse for p, hence p is a homotopy equivalence and $\gamma(p)$ is an isomorphism.

(iii) ==> (ii). By Theorem 1 $\gamma(p)$ an isomorphism ==> p is a homotopy equivalence and there is a map t: Y --> X with $pt \sim id_Y$ and $tp \sim id_X$. By the covering homotopy theorem we may assume that $pt = id_Y$. Let q: XxI --> X be a left homotopy from tp to id_X. Then the composite homotopy $q^{-1} \cdot tpq$: XxI' --> X from id_X to tp covers the composite homotopy $(pq)^{-1} \cdot pq$:XxI' --> Y from p to p. However proposition 1, §2 implies that $(pq)^{-1} \cdot (pq)$ is left homotopic to $p\sigma$: XxI --> Y, that is, there exists H: XxJ --> Y with $Hj_0 = p\sigma$ and $Hj_1 = (pq)^{-1} \cdot pq$ where XxJ, j_0, j_1, τ are as in (1), §2 with A replaced by X. By a covering homotopy argument which takes the form

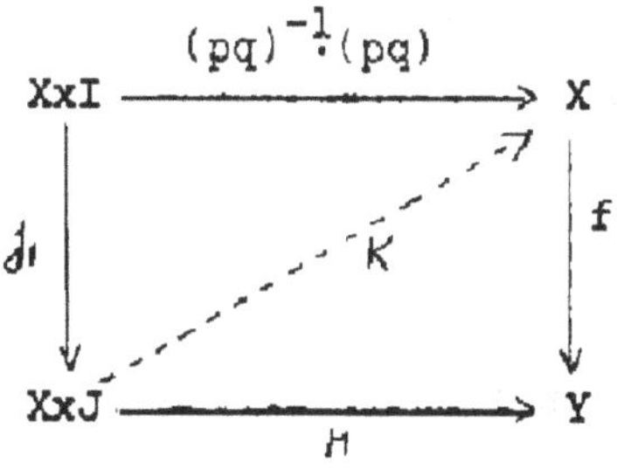

we obtain a left homotopy Kj_0: XxI --> Y from id_X to tp with $pKj_0 = p\sigma$ whose inverse is the desired homotopy h. QED.

<u>Definition</u>: A map $f: X \dashrightarrow Y$ is said to be a <u>retract</u> of a map $f': X' \dashrightarrow Y'$ if there is a diagram

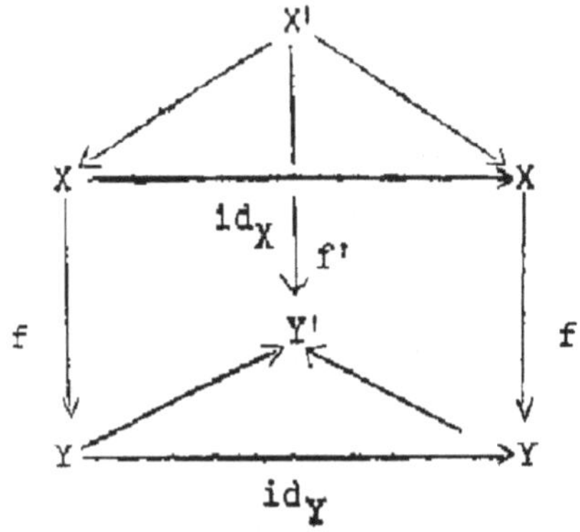

<u>Proposition 1</u>: Let $\underline{C}$ be a closed model category and let f be a map in $\underline{C}$. Then $\gamma(f)$ is an isomorphism iff f is a weak equivalence.

<u>Proof</u>: The direction $\Longleftarrow$ is the basic property of γ, so we suppose that $\gamma(f)$ is an isomorphism. By M5 and M2 we reduce to the case where f is a fibration in $\underline{C}_{cf}$ whence the result follows from the above lemma and M6(c).

<u>Proposition 2</u>: Let $\underline{C}$ be a model category. Then $\underline{C}$ is closed if and only if each of the classes of fibrations, cofibrations, and weak equivalences has the property that any retract of a member of the class is again a member.

<u>Proof</u>: ($\Longleftarrow$). Let $p: X \longrightarrow Y$ have the lifting property (1) whenever i is a trivial cofibration. By M2 we may factor p into $X \xrightarrow{i} Z \xrightarrow{u} Y$ where i is a

trivial cofibration and u is a fibration. By the property of p there is a dotted arrow s in

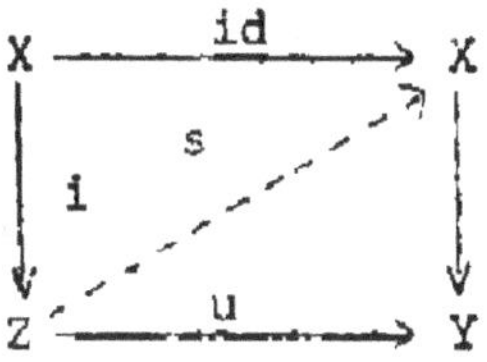

It follows that p is a retract of the fibration u and hence that p is a fibration. This proves a aince M1 gives the ==> direction of a, and the proof of b is similar. Suppose that f = uv as in c. Then by the above argument u is a retract of a trivial cofibration and hence by assumption is a weak equivalence. Similarly v is a weak equivalence and so f is also. This proves c since the implication ==> is contained in M2, M5, and M1. So C is closed.

(==>) It is immediate that a retract of a map with a lifting property of the kind in a) b) c) again has that lifting property. Thus the classes of fibrations and cofibrations, are closed under retracts. Let $\gamma: \underline{C} \longrightarrow \mathrm{Ho}\underline{C}$ be the canonical localization functor and suppose that f is a retract of a weak equivalence. Then $\gamma(f)$ is a retract of an isomorphism and hence is an isomorphism so f is a weak equivalence by proposition 1.

Chapter II. Examples of simplicial homotopy theories

Introduction

The first four sections of Chapter II give some examples of model categories. In §3 it is shown how the categories of topological spaces, simplicial groups, and simplicial sets form model categories, and in §4 this result is extended to the category $s\underline{A}$ of simplicial objects over a category $\underline{A}$, where $\underline{A}$ is a category closed under finite limits having sufficiently many projective objects and satisfying one of the following additional assumptions: (i) $\underline{A}$ has sufficiently many cogroup objects, (ii) $\underline{A}$ is closed under arbitrary inductive limits and has a set of small projective generators. The proofs for topological spaces, simplicial groups, and $s\underline{A}$ when $\underline{A}$ satisfies (ii) are similar and fairly simple, since every object in the model category is fibrant. For simplicial sets we were unable to find a really elementary proof; the argument given, which we think is the simplest, uses the classification theory of minimal fibrations [4]. It is possible to give another argument using the functor Ex^∞ of Kan [12] and a variant of this argument is used for $s\underline{A}$ in case (ii).

All of these categories are what we call <u>simplicial categories</u>, i.e. categories $\underline{C}$ endowed with a simplicial set "function complex" $\underline{Hom}_{\underline{C}}(X,Y)$ for each pair of objects X and Y satisfying suitable conditions. In §1 we define simplicial categories and the generalized path and cylinder functors $X,K \mapsto X \otimes K$, $Y,K \mapsto Y^K$, K a simplicial set, by the formulas

$$\underline{\mathrm{Hom}}_{\underline{S}}(K, \underline{\mathrm{Hom}}_{s\underline{A}}(X,Y)) = \underline{\mathrm{Hom}}_{s\underline{A}}(X \otimes K, Y) = \underline{\mathrm{Hom}}_{s\underline{A}}(X, Y^K)$$

where $\underline{S}$ is the category of simplicial sets. In §2 we define closed simplicial model category which is a category having the structures of a simplicial category and a closed model category compatibly related. All the examples of Chapter II are closed simplicial model categories; moreover for these model categories there are canonically adjoint path and cylinder functors, so that much of the work of the first chapter simplifies considerably (see [13]). However there are certain categories of differential graded algebras which do not seem to have natural simplicial structures but which are model categories, which is the main reason for the generality in Chapter I.

In §5 we show under suitable assumptions how homology and cohomology for model categories may be defined using abelian group objects and the abelianization functor. In particular we define cohomology groups of an object X with values in an abelian group object A of a model category $\underline{C}$. When $\underline{C}$ is the category of simplicial objects in a category $\underline{A}$ and X and A are constant simplicial objects, we show that these cohomology groups are equivalent to those obtained from suitable cotriples and Grothendieck sheaves. We also indicate how this cohomology gives a cohomology theory for arbitrary universal algebras coinciding up to a dimension shift with usual cohomology in the case of groups, and Lie algebras and associative algebras over a field.

In §6 we show that the category of simplicial modules over a simplicial ring forms a model category and use this to derive several Kunneth spectral sequences which will be used in later applications.

The present framework for homotopical algebra is not the most general that can be imagined. We have restricted ourselves to categories $\underline{A}$ closed under finite limits and having sufficiently many projective objects. The sheaf cohomology of Grothendieck is defined much more generally and Artin-Mazur [1] have shown in the case of the etale topology for preschemes that it gives rise to an analogue of ordinary homotopy theory using pro-objects in a homotopy category. It would also be nice to weaken the hypothesis that finite limits exist on a model category so the category of 1-connected pointed topological spaces would become a model category. Finally further generalization might eliminate the following inadequacy of this theory, that although derived functors may be defined for any category $\underline{A}$ with finite limits and enough projectives, the category $s\underline{A}$ does not form a model category without additional assumptions.

Chapter II. Examples of simplicial homotopy theories

§1. Simplicial Categories.

$\underline{S}$ will denote the category of (semi-) simplicial sets (see [7]).

Definition 1: A *simplicial category* is a category $\underline{C}$ endowed with the following structure:

(i) a functor $X,Y \mapsto \underline{\mathrm{Hom}}_{\underline{C}}(X,Y)$ from $\underline{C}^{o} \times \underline{C}$ to $\underline{S}$,

(ii) maps in $\underline{S}$

$$\underline{\mathrm{Hom}}_{\underline{C}}(X,Y) \times \underline{\mathrm{Hom}}_{\underline{C}}(Y,Z) \to \underline{\mathrm{Hom}}_{\underline{C}}(X,Z)$$

$$f,g \mapsto g \circ f$$

called *composition* defined for each triple X,Y,Z of objects of $\underline{C}$,

(iii) an isomorphism

$$\mathrm{Hom}_{\underline{C}}(X,Y) \xrightarrow{\sim} \underline{\mathrm{Hom}}_{\underline{C}}(X,Y)_{o}$$

$$u \mapsto \tilde{u}$$

of functors from $\underline{C}^{o} \times \underline{C}$ to (sets).

This structure is subject to the following two conditions:

(1) If $f \in \underline{\mathrm{Hom}}_{\underline{C}}(X,Y)_{n}$, $g \in \underline{\mathrm{Hom}}_{\underline{C}}(Y,Z)_{n}$ and $h \in \underline{\mathrm{Hom}}_{\underline{C}}(Z,W)_{n}$, then $(h \circ g) \circ f = h \circ (g \circ f)$.

(2) If $u \in \mathrm{Hom}_{\underline{C}}(X,Y)$ and $f \in \underline{\mathrm{Hom}}_{\underline{C}}(Y,Z)_n$, then $f \circ s_0^n \tilde{u} = \underline{\mathrm{Hom}}_{\underline{C}}(u,Z)_n(f)$. Also $s_0^n \tilde{u} \circ g = \underline{\mathrm{Hom}}_{\underline{C}}(W,u)_n(g)$ if $g \in \underline{\mathrm{Hom}}_{\underline{C}}(W,X)_n$.

Definition 2: Let $\underline{C}_1$ and $\underline{C}_2$ be simplicial categories. By a simplicial functor $F:\underline{C}_1 \to \underline{C}_2$ we mean a functor F from $\underline{C}_1$ to $\underline{C}_2$ together with maps $\underline{\mathrm{Hom}}_{\underline{C}_1}(X,Y) \to \underline{\mathrm{Hom}}_{\underline{C}_2}(FX,FY)$, denoted $f \mapsto F(f)$, such that $F(f \circ g) = F(f) \circ F(g)$ and $F(\tilde{u}) = \widetilde{F(u)}$.

Example: If X and Y are simplicial sets, let $\underline{\mathrm{Hom}}_{\underline{S}}(X,Y)$ or simply $\underline{\mathrm{Hom}}(X,Y)$ be the "function complex" simplicial set of maps from X to Y. There is a canonical "evaluation map"

$$\text{(1)} \qquad ev: X \times \underline{\mathrm{Hom}}(X,Y) \to Y$$

giving rise to isomorphisms

$$\text{(2)} \qquad \mathrm{Hom}(K, \underline{\mathrm{Hom}}(X,Y)) \xrightarrow[\sim]{\#} \mathrm{Hom}(X \times K, Y)$$

for all $K \in \mathrm{Ob}\,\underline{S}$, where $\#(u) = ev \circ (id_X \times u)$. The map

$$X \times \underline{\mathrm{Hom}}(X,Y) \times \underline{\mathrm{Hom}}(Y,Z) \xrightarrow{ev \times id} Y \times \underline{\mathrm{Hom}}(Y,Z) \xrightarrow{ev} Z$$

thereby determines a composition map (ii), while taking $K = \Delta(0)$, the final object of $\underline{S}$, in (2) yields an isomorphism (iii). It is easily seen that $\underline{S}$ is a simplicial category.

If X is a fixed object of $\underline{S}$, then the functor $Y \mapsto \underline{\mathrm{Hom}}(X,Y)$ is a simplicial functor h^X, where $h^X: \underline{\mathrm{Hom}}(Y,Z) \to \underline{\mathrm{Hom}}(\underline{\mathrm{Hom}}(X,Y), \underline{\mathrm{Hom}}(X,Z))$ is given by $\#(h^X) = $ composition.

In the following $\underline{C}$ denotes a simplicial category. When convenient we will identify $\mathrm{Hom}_{\underline{C}}(X,Y)_0$ with $\underline{\mathrm{Hom}}_{\underline{C}}(X,Y)_0$ and drop the "$\sim$" notation. Also we will often write $\underline{\mathrm{Hom}}(X,Y)$ instead of $\underline{\mathrm{Hom}}_{\underline{C}}(X,Y)$.

__Definition 3__: Let $X \in \mathrm{Ob}\underline{C}$ and $K \in \mathrm{Ob}\underline{S}$. By $X \otimes K$ we shall denote an object of $\underline{C}$ with a distinguished map $\alpha: K \to \underline{\mathrm{Hom}}_{\underline{C}}(X, X\otimes K)$ such that

(3) $$\varphi: \underline{\mathrm{Hom}}_{\underline{C}}(X\otimes K, Y) \cong \underline{\mathrm{Hom}}_{\underline{S}}(K, \underline{\mathrm{Hom}}_{\underline{C}}(X,Y))$$

for all $Y \in \mathrm{Ob}\underline{C}$, where $\#(\varphi)$ is the map

$$K \times \underline{\mathrm{Hom}}(X\otimes K, Y) \xrightarrow{\alpha \times id} \underline{\mathrm{Hom}}(X, X\otimes K) \times \underline{\mathrm{Hom}}(X\otimes K, Y) \xrightarrow{\circ} \underline{\mathrm{Hom}}(X,Y) .$$

By X^K we denote an object of $\underline{C}$ with a map $\beta: K \to \underline{\mathrm{Hom}}_{\underline{C}}(X^K, X)$ such that

(4) $$\psi: \underline{\mathrm{Hom}}_{\underline{C}}(Y, X^K) \cong \underline{\mathrm{Hom}}_{\underline{S}}(K, \underline{\mathrm{Hom}}_{\underline{C}}(Y,X))$$

for all $Y \in \mathrm{Ob}\underline{C}$, where $\#(\psi)$ is the composition

$$K \times \underline{\mathrm{Hom}}(Y, X^K) \xrightarrow{(pr_2, \beta pr_1)} \underline{\mathrm{Hom}}(Y, X^K) \times \underline{\mathrm{Hom}}(X^K, X) \xrightarrow{\circ} \underline{\mathrm{Hom}}(Y,X) .$$

__Example__: If $\underline{C} = \underline{S}$, then $X \times K$ together with the map $\alpha: K \to \underline{\mathrm{Hom}}(X, X\times K)$ such that $\#(\alpha) = id_{X\times K}$ is an object $X \otimes K$. $\underline{\mathrm{Hom}}(K,X)$ with the map $\beta: K \to \underline{\mathrm{Hom}}(\underline{\mathrm{Hom}}(K,X),X)$ such that $\#\beta$ = the composition $\underline{\mathrm{Hom}}(K,X) \times K \xrightarrow{(pr_2, pr_1)} K \times \underline{\mathrm{Hom}}(K,X) \xrightarrow{ev} X$ is an object X^K .

<u>Proposition 1</u>: If $X \in \mathrm{Ob}\underline{C}$ and $K,L \in \mathrm{Ob}\underline{S}$, then there are canonical isomorphisms

$$(5) \qquad X \otimes (K \times L) \simeq (X \otimes K) \otimes L \qquad\qquad (X^K)^L \simeq X^{K \times L}$$

when all the objects are defined.

<u>Proof</u>: $\mathrm{Hom}(X \otimes (K \times L), Y) = \mathrm{Hom}(K \times L, \underline{\mathrm{Hom}}(X,Y)) =$ $\mathrm{Hom}(L, \underline{\mathrm{Hom}}(K, \underline{\mathrm{Hom}}(X,Y))) = \mathrm{Hom}(L, \underline{\mathrm{Hom}}(X \otimes K, Y)) = \mathrm{Hom}((X \otimes K) \otimes L, Y)$. This yields the first isomorphism; the second is proved similarly.

<u>Remark 1</u>: The degree-0 part of (3) yields the formula

$$(6) \qquad \mathrm{Hom}_{\underline{C}}(X \otimes K, Y) = \mathrm{Hom}_{\underline{S}}(K, \underline{\mathrm{Hom}}_{\underline{C}}(X,Y)) \ .$$

The difference between (6) and (3) is roughly the first isomorphism of (5) as one sees by analyzing the proof of (5). In practice (see Prop. 2 below) one defines an operation $X \otimes K$ satisfying (6) and (5) and then proves (3) by inverting the proof of (5).

2: The objects $X \otimes K$ and X^K have the following interpretation whose details we leave to the reader. The functor $Y \mapsto \underline{\mathrm{Hom}}(X,Y)$ is a simplicial functor h^X from $\underline{C}$ to $\underline{S}$ in a natural way. Call a simplicial functor $F: \underline{C} \to \underline{S}$ representable if it is isomorphic to h^X for some $X \in \mathrm{Ob}\underline{C}$. (Yoneda's lemma holds: $\underline{\mathrm{Hom}}_{\underline{F}}(h^X, F) = F(X)$, where $\underline{F}$ is the <u>simplicial</u> category of simplicial functors from $\underline{C}$ to S .) Then $X \otimes K$ represents the simplicial functor $Y \mapsto \underline{\mathrm{Hom}}(K, \underline{\mathrm{Hom}}(X,Y))$.

Let $\pi_0(K)$ be the set of components of the simplicial set K so that we have adjoint functors

(7) $$\mathrm{Hom}_{\underline{S}}(K, K(S,0)) \simeq \mathrm{Hom}_{(\mathrm{sets})}(\pi_0(K), S)$$

where if S is a set $K(S,0)$ denotes the constant simplicial set which is S in each dimension and has all simplicial operators $= \mathrm{id}_S$. If $x, y \in K_0$ we say that x is <u>strictly homotopic</u> to y if there is a z in K_1 with $d_1 z = x$ and $d_0 z = y$ and that x is <u>homotopic</u> to y if x and y are equivalent with respect to the equivalence relation generated by the relation "is strictly homotopic to". $\pi_0(K)$ is the quotient of K_0 by the relation "is homotopic to" and hence

(8) $$\pi_0(K \times L) \cong \pi_0(K) \times \pi_0(L)$$

Let J denote a generalized unit interval, that is, a simplicial set which is a string of copies of $\Delta(1)$ joined end to end. Let $\{0\} \subset J$ and $\{1\} \subset J$ be the subcomplexes generated by the first and last vertices of J. A typical J may be pictured

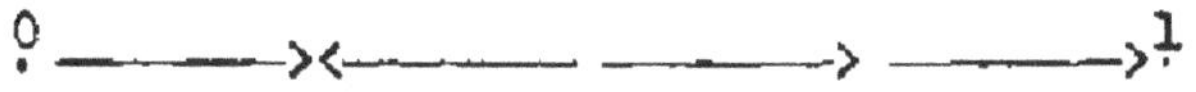

and it is clear that two simplices x and y of K are homotopic if there exists a generalized unit interval J and a map $u: J \to K$ with $u(0) = x$ and $u(1) = y$.

Definition 4: Let X,Y be two objects of $\underline{C}$ and f,g two maps from X to Y. We say that f is strictly homotopic (resp. homotopic) to g if this is the case when f and g are regarded as 0-simplices of $\underline{\mathrm{Hom}}(X,Y)$. By a strict homotopy (resp. homotopy) from f to g we mean an element $h \in \underline{\mathrm{Hom}}(X,Y)_1$ with $d_1h = f$ and $d_0h = g$ (resp. a map $u:J \to \underline{\mathrm{Hom}}(X,Y)$ with $u(0) = f$ and $u(1) = g$). Let $\pi_0(X,Y) = \pi_0\underline{\mathrm{Hom}}(X,Y)$ be the homotopy classes of maps from X to Y. We define the category $\pi_0\underline{C}$ to be the category with the same objects as $\underline{C}$, with $\mathrm{Hom}_{\pi_0\underline{C}}(X,Y) = \pi_0(X,Y)$, and with composition induced from the composition in $\underline{C}$ (this is legitimate by (8)).

When objects $X \otimes K$ and X^K exist in $\underline{C}$, then a homotopy from f to g is the same as a map $H:X\otimes J \to Y$ with $Hi_0 = f$, $Hi_1 = g$. Here J is a generalized unit interval and $i_e:X \to X\otimes J$ denotes the map induced by the 0-simplex e of J where $e = 0$ or 1. The homotopy may also be identified with a map $H':Y \to Y^J$ with $j_0H' = f$ and $j_1H' = g$ where $j_e:X^J \to X$ is induced by $e \in J_0$. The reader will note that we have changed notation from ∂, d of Ch. I to i,j. This is because d_0 corresponds to i_1. However we will retain the notation $s:X \to X^J$ and $\sigma:X\otimes J \to X$ to denote the constant homotopy of id_X. These are the maps induced by the unique map $J \to \Delta(0)$.

Let $\underline{A}$ be a category and let $s\underline{A}$ be the category of simplicial objects over $\underline{A}$, that is, contravariant functors $\Delta \to \underline{A}$, where Δ is the category having for objects the ordered sets $[n] = \{0,1,\ldots,n\}$ for each integer $n \geq 0$, and where a map $\varphi:[p] \to [q]$ in Δ is a (weakly) monotone map. If X is an object of $s\underline{A}$, we write X_n instead of $X[n]$ and φ_X^* (or simply φ^*) for $X(\varphi)$ when φ is a map in Δ. If X,Y are objects of $s\underline{A}$ and if K is a simplicial set, then ([@],[13]) a map $f:X\times K \to Y$ is defined to be a collection of maps $f(\sigma):X_q \to Y_q$, one for each $q \geq 0$ and $\sigma \in K_q$, such that $\varphi_Y^* f(\sigma) = f(\varphi_K^*\sigma)\varphi_X^*$ for any map φ in Δ. $X\times K$ is not to be understood as an object of $s\underline{A}$ and f is not a morphism in a category. Letting $\mathrm{Map}(X\times K,Y)$ be the set of maps $f:X\times K \to Y$ we obtain a functor $(s\underline{A})^{\circ} \times \underline{S}^{\circ} \times (s\underline{A}) \to (\text{sets})$ and hence a functor $X,Y \mapsto \underline{\mathrm{Hom}}_{s\underline{A}}(X,Y)$ from $(s\underline{A})^{\circ} \times (s\underline{A})$ to $\underline{S}$ given by $\underline{\mathrm{Hom}}_{s\underline{A}}(X,Y)_n = \mathrm{Map}(X\times\Delta(n),Y)$ with simplicial operator $\varphi^* = \mathrm{Map}(X\times\tilde{\varphi},Y)$. Here $\Delta(n)$ is the "standard n-simplex" simplicial set, which is the functor $\Delta^{\circ} \to (\text{sets})$ represented by $[n]$, and for any simplicial set K and $\sigma \in K_n$ we let $\tilde{\sigma}:\Delta(n) \to K$ be the unique map in $\underline{S}$ with $\tilde{\sigma}(\mathrm{id}_{[n]}) = \sigma$.

If $X,Y,Z \in \mathrm{Ob}\, s\underline{A}$ and K is a simplicial set, then we may define the composition $g\circ f$ of two maps $f:X\times K \to Y$ and $g:Y\times K \to Z$ by $(g\circ f)(\sigma) = g(\sigma)f(\sigma)$. This yields a composition operation as in (ii) Def. 1, and (iii) comes from the fact that $\Delta(0)_q$ consists of exactly one element for each q. It is clear that $s\underline{A}$ thereby becomes a simplicial category. Also if the functor $F:\underline{A} \to \underline{B}$ is extended degree-wise to $sF:s\underline{A} \to s\underline{B}$, then sF is a simplicial functor where if $f:X\times K \to Y$ we let $(sF)(f):F(X)\times K \to F(Y)$ be given by $[(sF)(f)](\sigma) = F(f(\sigma))$.

Recall that a simplicial set is said to be <u>finite</u> if it has only finitely many non-degenerate simplices. A finite simplicial set is always a simplicial finite set, i.e. a simplicial object over the category of finite sets, but not conversely.

<u>Proposition 2</u>: Let $\underline{A}$ be a category and let X be a simplicial object over $\underline{A}$. If $\underline{A}$ is closed under (finite) direct sums, then $X\underline{\otimes}K$ exists in $s\underline{A}$ for every simplicial (finite) set K . If $\underline{A}$ is closed under (finite) projective limits, then X^K exists for every (finite) simplicial set K .

<u>Proof</u>: Let $(X\underline{\otimes}K)_n = \bigvee_{\sigma\in K_n} X_n$ with $\varphi^*_{X\underline{\otimes}K} = \sum_\sigma in_{(\varphi^*_K\sigma)}\varphi^*_X$.

Here $\bigvee_{i\in I} X_i$ denotes the direct sum of an indexed family $\{X_i; i \in I\}$ of objects of $\underline{A}$, $in_i:X_i \to \bigvee X_i$ is the injection of the ith component, and $\Sigma f_i:\bigvee X_i \to Y$ is the unique map with $(\Sigma f_i)in_j = f_j$ for all $j \in I$ if $\{f_i:X_i \to Y , i \in I\}$ is a family of maps in $\underline{A}$. These direct sums exist by the assumptions on $\underline{A}$ and K . Let $\xi:X\times K \to X\underline{\otimes}K$ be given by $\xi(\sigma) = in_\sigma$. Finally let $ev':X\times\underline{Hom}_{s\underline{A}}(X,Y) \to Y$ be given by $ev'(f_n) = f_n(id_{[n]}):X_n \to Y_n$. Then there are isomorphisms

$$Hom_{\underline{S}}(K,\underline{Hom}_{s\underline{A}}(X,Y)) \xrightarrow[\sim]{\#'} Map(X\times K,Y) \xleftarrow[\sim]{\xi^*} Hom_{s\underline{A}}(X\underline{\otimes}K,Y)$$

where $\#'$ is induced by ev' and ξ^* by ξ . Letting

$\# = (\xi^*)^{-1} \circ (\#')$ it is clear that $\#$ is functorial as X, Y run over $s\underline{A}$ and K varies over the category of simplicial (finite) sets. If L is another simplicial (finite) set, then there is a canonical isomorphism

$$\theta : X \underline{\otimes} (K \times L) \simeq (X \underline{\otimes} K) \underline{\otimes} L$$

given by $\theta_n = \sum_{(\sigma,\tau) \in (K \times L)_n} in_\tau in_\sigma$.

Now if $\alpha : K \to \underline{Hom}(X, X \underline{\otimes} K)$ is given by $\#(\alpha) = id_{X \underline{\otimes} K}$, then $X \underline{\otimes} K$ with α is an object $X \otimes K$. In effect letting φ be the map (3) determined by α , we have the diagram

$$\begin{array}{ccc}
Hom(L, \underline{Hom}(X \underline{\otimes} K, Y)) & \xrightarrow{\varphi^*} & Hom(L, \underline{Hom}(K, \underline{Hom}(X, Y))) \\
 & & \downarrow \# \wr \\
\# \wr \downarrow & & Hom(K \times L, \underline{Hom}(X, Y)) \\
 & & \downarrow \# \wr \\
Hom((X \underline{\otimes} K) \underline{\otimes} L, Y) & \underset{\sim}{\xleftarrow{\theta^*}} & Hom(X \underline{\otimes} (K \times L), Y)
\end{array}$$

which may be shown to be commutative by a straightforward analysis of the definitions. Taking $L = \Delta(n)$ for each n we see that φ is an isomorphism and hence the first part of the proposition is proved.

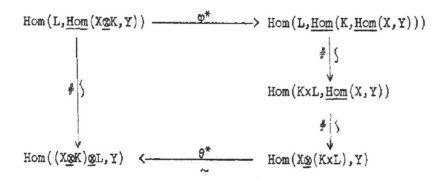

Let $\underline{A}'$ be the category of functors $\underline{A}^\circ \to$ (sets) and let $X \mapsto hX$ be the canonical fully faithful functor (this forces us to leave the haven of our universe). Denoting the degree-wise extension of h by $h: s\underline{A} \to s\underline{A}'$, one sees that

$$\mathrm{Map}_{s\underline{A}}(X,Y) \simeq \mathrm{Map}_{s\underline{A}'}(hX \times K, hY) \quad , \quad \text{so}$$

$$(9) \qquad \underline{\mathrm{Hom}}_{s\underline{A}}(X,Y) \xrightarrow{\approx} \underline{\mathrm{Hom}}_{s\underline{A}'}(hX,hY)$$

and h is a "fully faithful" simplicial functor. Now if $F \in \mathrm{Ob}\, s\underline{A}'$, then F^K exists and is given by $F^K(A) = F(A)^K$ for all $A \in \mathrm{Ob}\,\underline{A}$, where we have identified $s\underline{A}'$ with the category of functors $\underline{A}^o \to \underline{S}$ in the natural way. One sees immediately from (9) that X^K exists if and only if $(hX)^K$ is isomorphic to hZ for some $Z \in \mathrm{Ob}\, s\underline{A}$, or equivalently if $[(hX)^K]_n$ is a representable functor for each n. There is a cokernel diagram in $\underline{S}$

$$\bigvee_{j \in J} \Delta(q_j) \rightrightarrows \bigvee_{i \in I} \Delta(p_i) \to K \times \Delta(n)$$

where if K is finite so is $K \times \Delta(n)$ and hence I and J are finite sets. But the functor

$$A \mapsto [(hX)^K]_n(A) = \mathrm{Hom}_{\underline{S}}(K \times \Delta(n)\ ,\ hX(A))$$

$$= \mathrm{Ker}\{\prod_I hX_{p_i}(A) \rightrightarrows \prod_J hX_{q_j}(A)\}$$

$$= h\mathrm{Ker}\{\prod_I X_{p_i} \rightrightarrows \prod_J X_{q_j}\}(A)$$

is representable by the assumptions made on $\underline{A}$. Q.E.D.

Corollary: If $F:\underline{A} \to \underline{B}$ commutes with (finite) direct sums (resp. projective limits), then $F(X) \otimes K \cong F(X \otimes K)$ for all $X \in \mathrm{Ob}\, s\underline{A}$ and simplicial (finite) sets K (resp. $F(X^K) \cong F(X)^K$ for all X and (finite) simplicial sets K).

This is immediate from the formulas for $X \otimes K$ and X^K obtained in the proof of Prop. 2.

Remark: The corollary implies that if G is a simplicial group then the underlying simplicial set of G^K is (underlying simplicial set of $G)^K$, and similarly for any other algebraic species.

§2. Closed simplicial model categories.

$\dot{\Delta}(n)$ for $n \geq 0$ (resp. $V(n,k)$ for $0 \leq k \leq n > 0$) denotes the simplicial subset of $\Delta(n)$ which is the union of the images of the faces $\partial_i : \Delta(n-1) \to \Delta(n)$ for $0 \leq i \leq n$ (resp. $0 \leq i \leq n$, $i \neq k$). $\dot{\Delta}(0) = \emptyset$ the initial object in $\underline{S}$. In the following RLP (resp. LLP) stands for right (resp. left) lifting property (Ch. I, §5).

Proposition 1: The following are equivalent for a map f in $\underline{S}$:

(i) f has the RLP with respect to $\dot{\Delta}(n) \hookrightarrow \Delta(n)$ for all n

(ii) f " " " " " " any injective (i.e. injective in each degree) map of simplicial sets.

This follows immediately from the skeletal decomposition of an injective map (see [7], Ch. II, 3.8). The following is proved in [7], Ch. IV, §2.1. $\{e\} \subset \Delta(1)$ denotes the subcomplex consisting of the degeneracies of the vertex e, where $e = 0,1$.

Proposition 2: The following are equivalent for a map f in $\underline{S}$.

(i) f has the RLP with respect to $V(n,k) \hookrightarrow \Delta(n)$ for $0 \leq k \leq n > 0$.

(ii) f " " " " " " $\dot{\Delta}(n) \times \Delta(1) \cup \Delta(n) \times \{e\} \hookrightarrow \Delta(n) \times \Delta(1)$ for $n \geq 0$ and $e = 0,1$.

(iii) f has the RLP with respect to $L\times\Delta(1) \cup K\times\{e\} \hookrightarrow K\times\Delta(1)$ for all injective maps $L \hookrightarrow K$ in $\underline{S}$ and $e = 0,1$.

Definition 1: A map of simplicial sets will be called a trivial fibration (resp. fibration) if it satisfies the equivalent conditions of Proposition 1 (resp. Proposition 2).

Thus a fibration is a fiber map in the sense of Kan. It is easy to see that a trivial fibration is a fibration whose fibers are contractible.

Definition 2: By a closed simplicial model category we mean a closed model category $\underline{C}$ which is also a simplicial category satisfying the following two conditions:

SM0: If $X \in \mathrm{Ob}\underline{C}$, then the objects $X\otimes K$ and X^K exist for any finite simplicial set K.

SM7: If $i:A \to B$ is a cofibration and $p:X \to Y$ is a fibration, then

$$(1) \qquad \underline{\mathrm{Hom}}(B,X) \xrightarrow{(i^*,p_*)} \underline{\mathrm{Hom}}(A,X) \times_{\underline{\mathrm{Hom}}(A,Y)} \underline{\mathrm{Hom}}(B,Y)$$

is a fibration of simplicial sets, which is trivial if either i or p is trivial.

Convention: It will be convenient to use the notation $\underline{\mathrm{Hom}}(i,p)$ for the target of the map (1).

Proposition 3: Suppose that $\underline{C}$ is a simplicial category satisfying M0 and SM0 with four distinguished classes of maps--fibrations, cofibrations, trivial fibrations, and trivial cofibrations--such that the first and fourth (resp. second and third) determine each other by lifting properties as in M6 (a) and (b). (This holds in particular if $\underline{C}$ is a closed simplicial model category.) Then SM7 is equivalent separately to each of the following:

SM7(a). If $X \to Y$ is a fibration (resp. trivial fibration), then $X^{\Delta(n)} \to X^{\dot{\Delta}(n)} \times_{Y^{\dot{\Delta}(n)}} Y^{\Delta(n)}$ is a fibration (resp. trivial fibration) and $X^{\Delta(1)} \to X^{\{e\}} \times_{Y^{\{e\}}} Y^{\Delta(1)}$ is a trivial fibration for $e = 0,1$.

SM7(b). If $A \to B$ is a cofibration (resp. trivial cofibration), then $A \otimes \Delta(n) \vee_{A \otimes \dot{\Delta}(n)} B \otimes \dot{\Delta}(n) \to B \otimes \Delta(n)$ is a cofibration (resp. trivial cofibration, and $A \otimes \Delta(1) \vee_{A \otimes \{e\}} B \otimes \{e\} \to B \otimes \Delta(1)$ is a trivial cofibration for $e = 0,1$.

Proof: To show that $X^K \to X^L \times_{Y^L} Y^K$ is a fibration where $L \to K$ is a map of finite simplicial sets, it suffices to show that it has the RLP with respect to any trivial cofibration $A \to B$. By the definition of the object X^K this is equivalent to showing that $\underline{\mathrm{Hom}}(B,X) \to \underline{\mathrm{Hom}}(A,X) \times_{\underline{\mathrm{Hom}}(A,Y)} \mathrm{Hom}(B,Y)$ has the

RLP with respect to $L \to K$. Manipulating in this way one proves the proposition.

Remark: It is clear that SM7(a) holds for the fibrations and trivial fibrations in $\underline{S}$.

For the rest of this section $\underline{C}$ denotes a closed simplicial model category. We shall be concerned with relating the simplicial homotopy structure of $\underline{C}$ with the left and right homotopy structure of Ch.I. Let $f \overset{s}{\sim}{}^{s}\, g$ (resp. $f \overset{s}{\sim} g$) mean f is strictly (simplicially) homotopic (resp. (simplicially) homotopic) to g. The following is the covering homotopy extension theorem for simplicial homotopies. It should be noted how much stronger it is than the Cor. of Lemma 2 and Lemma 7 of §1, Ch.I.

Proposition 4: Let $i:A \to B$ be a cofibration and let $p:X \to Y$ be a fibration. Let $h:A\otimes J \to X$ and $h:B\otimes J \to Y$ be simplicial homotopies compatible with i and p in the sense that $pk = h(i\otimes id_J)$.

(1) If $\theta:B \to X$ satisfies $p\theta = hj_0$, $\theta i = ki_0$, then there is a homotopy $H:B\otimes J \to X$ with $Hi_0 = \theta$, $pH = h$, and $H(i\otimes id_J) = k$.

(2) If either i or p is trivial and if $\theta_e: B \to X$ satisfies $p\theta_e = hi_e$, $\theta i = ki_e$, $e = 0,1$, then there is a homotopy $H:B\otimes J \to X$ with $Hi_e = \theta_e$, $e = 0,1$, $pH = h$, and $H(i\otimes id_J) = k$.

Proof: This follows immediately from SM7 by an induction on the length of J.

Corollary: Let $i: A \to B$ be a cofibration of fibrant objects. Then i is trivial iff i is a strong deformation retract map (i.e. there exists $r: B \to A$, $h: B \otimes \Delta(1) \to B$ with $ri = id_A$, $h_0 = id_B$, $h_1 = ir$, $h(i \otimes \Delta(1)) = i\sigma$). Dually if $p: X \to Y$ is a fibration of cofibrant objects, then p is trivial iff there are maps $s: Y \to X$, $h: X \otimes \Delta(1) \to X$ with $ps = id_Y$, $h_0 = id_X$, $h_1 = sp$, $ph = \sigma(p \otimes \Delta(1))$.

Proof: ($\Longrightarrow$) r and h may be obtained by lifting successively in

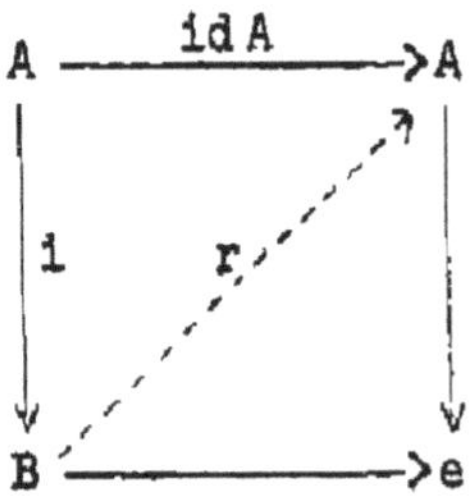

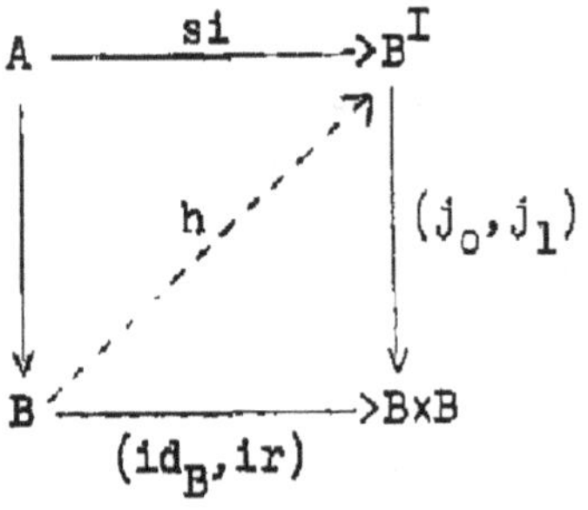

($\Longleftarrow$) is clear from Proposition 4.

Proposition 5: (1) If $f, g: X \rightrightarrows Y$ are two maps in $\underline{C}$, then $f \overset{\underline{s}}{\sim} g \Longrightarrow f \overset{\ell}{\sim} g$ and $f \overset{r}{\sim} g$. If X is cofibrant and Y is fibrant, then the strict simplicial, simplicial, left, and right homotopy relations on $\mathrm{Hom}(X,Y)$ coincide and are equivalence relations.

(2) The conclusions of Theorem 1, Ch.I, §1 remain valid if $\pi\underline{C}_c$, $\pi\underline{C}_f$, and $\pi\underline{C}_{cf}$ are replaced by $\pi_o\underline{C}_c$, $\pi_o\underline{C}_f$, and $\pi_o\underline{C}_{cf}$, respectively.

Proof: (2). The inclusion $\{0\} \subset J$ has the LLP with respect to fibrations in $\underline{S}$, hence if X is cofibrant one finds, as in the proof of Prop. 3(b), that $i_o: X \to X\otimes J$ is a trivial cofibration. By M5 the map $\sigma: X\otimes J \to X$ is a weak equivalence. Also by Prop 3(b) $X \vee X \xrightarrow{i_o+i_1} X\otimes J$ is a cofibration and so $X\otimes J$ is a cylinder object for J. It follows as in the proof of Lemma 8 (ii), Ch.I, §1, that if $f,g: X \rightrightarrows Y$ are two maps in $\underline{C}_c$ and $f \overset{s}{\simeq} g$, then $\gamma_c(f) = \gamma_c(g)$ and hence γ_c induces $\overline{\gamma}_c: \pi_o\underline{C}_c \to \mathrm{Ho}\underline{C}_c$. Similarly one shows that $\overline{\gamma}$, $\overline{\gamma}_f$ as in Theorem 1, Ch.I, §1 exist with π replaced by π_o. Next note that the "quasi-" functors $X \mapsto Q(X)$ and $X \mapsto R(X)$ of the proof of this theorem yield functors $\overline{Q}: \pi_o\underline{C} \to \pi_o\underline{C}_c$, $\overline{R}: \pi_o\underline{C} \to \pi_o\underline{C}_f$ in virtue of Prop. 4 (2) above. The rest of the proof of Theorem 1 goes through without change so (2) follows.

(1) The quasi-inverse of $\overline{\gamma}: \pi_o\underline{C}_{cf} \to \mathrm{Ho}\underline{C}$ constructed in the proof of Theorem 1 is induced by $\overline{RQ}: \underline{C} \to \pi_o\underline{C}_{cf}$. But we have just seen that $f \overset{s}{\simeq} g \Longrightarrow RQ(f) \overset{s}{\simeq} RQ(g)$ and therefore we conclude that $f \overset{s}{\simeq} g \Longrightarrow \gamma(f) = \gamma(g)$. Now if J is a generalized unit interval, there is a canonical homotopy $h: J\times J \to J$ with

$h(id_J \times \tilde{0}) = id_J$ and $h(id_J \times \tilde{1}) = id_J$ where $\tilde{e}: \Delta(0) \to J$ is the map with $\tilde{e}(id_{[0]}) = e$, $e = 0,1$ and σ is the unique map $J \to \Delta(0)$. This homotopy in a representative case may be pictured

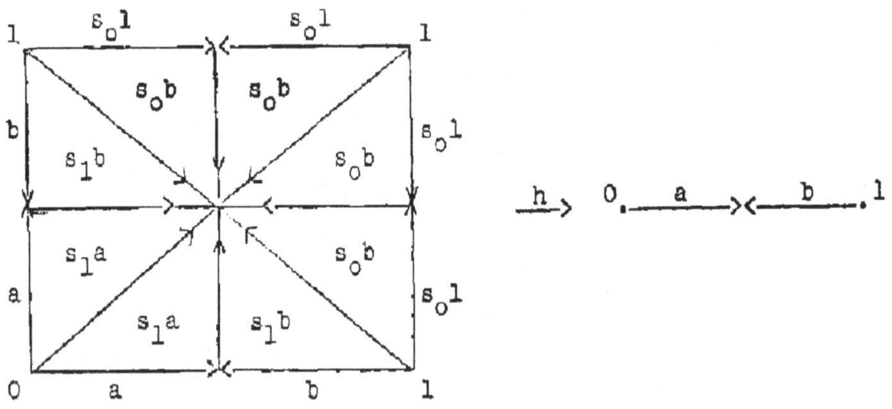

where the arrows denote the direction of each 1 simplex of $J \times J$ and where a simplex of $J \times J$ labelled $s_0 a$ goes to $s_0 a$ in J under h . Consequently if X is any object of $\underline{C}$, $\sigma : X \otimes J \to X$ is a simplicial homotopy equivalence and therefore $\gamma(\sigma)$ is an isomorphism. By Ch.I, §5, Prop.1, σ is a weak equivalence and therefore $f \overset{s}{\simeq} g \Longrightarrow f \overset{\ell}{\simeq} g$. Similarly $X \overset{s}{\to} X^J$ is a weak equivalence for all X in $\underline{C}$ so $f \overset{s}{\simeq} g \Longrightarrow f \overset{r}{\simeq} g$; thus this first part of (1) is proved. The last assertion follows from Lemma 1, §2, Ch.I which shows when X is cofibrant and Y is fibrant the cylinder object $X \otimes \Delta(1)$ (see proof of (2) above) may be used to represent any left homotopy from f to g and from

Lemma 4, §1, Ch.I. Q.E.D.

Remark: Proposition 5 shows that the simplicial homotopy relation on Hom(X,Y) is finer than either left or right homotopy, but when X is cofibrant and Y is fibrant the three relations coincide. One may compare the constructions of §2 and 3 of Ch.I with the corresponding well-known simplicial constructions and show that the resulting structure on Ho$\underline{C}$ is the same. Thus the fundamental groupoid of the Kan complex $\underline{\text{Hom}}(X,Y)$ coincides with the one constructed in §2, Ch.I, and if $E \to B$ is a fibration in $\underline{C}_f$ where $\underline{C}$ is pointed, then the long exact sequence of homotopy groups arising from the fibration $\underline{\text{Hom}}(A,E) \to \underline{\text{Hom}}(A,B)$ (SM7 when $A \in \text{Ob}\underline{C}_c$) coincides with that of §3, Ch.I.

Proposition 6: If $\underline{C}$ is a closed simplicial model category, then in a natural way so are the dual $\underline{C}^o$ and the category $\underline{C}/X$ of objects of $\underline{C}$ over a fixed object X .

Proof: The assertion about $\underline{C}^o$ is trivial. If A and B are two objects of $\underline{C}/X$, we let $\underline{\text{Hom}}_{\underline{C}/X}(A,B)$ be the subcomplex of $\underline{\text{Hom}}_{\underline{C}}(A,B)$ consisting of elements f_n of dimension n with $(s_o^n v)\circ f = s_o^n u$, where $u:A \to X$ and $v:B \to X$ are the structural maps. With the induced composition $\underline{C}/X$ becomes a simplicial category closed under finite limits. If K is a finite simplicial set, then the object $(A \xrightarrow{u} X)\otimes K$ in $\underline{C}/X$ is the map $A\otimes K \xrightarrow{\sigma(u\otimes id)} X$, where $\sigma:X\otimes K \to X$ is the map corresponding to the map $K \to \underline{\mathrm{Hom}}(X,X)$ sending all elements of K to degeneracies of id_X. The object $(A \overset{u}{\to} X)^K$ in $\underline{C}/X$ is the map $pr_2 : A^K \times_{X^K} X$, whose source is the fiber product of u^K and the map $s: X \to X^K$ corresponding to σ. Thus $\underline{C}/X$ satisfies SM0.

A map in $\underline{C}/X$ will be called a fibration, cofibration or weak equivalence if it is so in $\underline{C}$. Axioms M2 and M5 are clear. If $i: A \to A'$ and $p: B' \to B$ are maps in $\underline{C}/X$, then the map $\mathrm{Hom}_{\underline{C}/X}(A',B') \to \underline{\mathrm{Hom}}_{\underline{C}/X}(i,p)$ is the base extension by the structural map $\Delta(0) \to \underline{\mathrm{Hom}}_{\underline{C}}(A',X)$ of the map $\mathrm{Hom}_{\underline{C}}(A',B') \to \mathrm{Hom}_{\underline{C}}(i,p)$. Hence SM7 holds, hence also M1. To obtain M6 argue as follows: Supposing a map f in $\underline{C}/X$ has the LLP with respect to the fibrations in $\underline{C}/X$, factor $f = pi$ where i is a trivial cofibration and p is a fibration in $\underline{C}/X$; then f is a retract of i hence is a trivial cofibration in $\underline{C}$ and hence in $\underline{C}/X$. The other cases of M6 are similar. Q.E.D.

§3. Spaces, simplicial groups, and simplicial sets.

Let $\underline{T}$ be the category of topological spaces and continuous maps. If X and Y are spaces, define the function complex $\underline{Hom}(X,Y)$ by $\underline{Hom}(X,Y)_n = Hom(X\times|\Delta(n)|,Y)$ with natural simplicial operations, where $|\ |$ denotes geometric realization. If $f \in \underline{Hom}(X,Y)_n$ and $g \in \underline{Hom}(Y,Z)_n$, let $g\circ f$ be the composite map

$$X\times|\Delta(n)| \xrightarrow{id\times\Delta} X\times|\Delta(n)| \times |\Delta(n)| \xrightarrow{f\times id} Y\times|\Delta(n)| \xrightarrow{g} Z .$$

$\underline{T}$ thereby becomes a simplicial category where $X\otimes K = X\times|K|$ and X^K = the function space $X^{|K|}$.

A map $f:X \to Y$ in $\underline{T}$ will be called a fibration if it is a fiber map in the sense of Serre and a weak equivalence if it is a weak homotopy equivalence (i.e. $\pi_q(X,x) \xrightarrow{\sim} \pi_q(Y,fx)$ for all $x \in X$ and $q \geq 0$). Finally a map will be called a cofibration if it has the LLP with respect to all trivial fibrations.

Theorem 1: With these definitions the category $\underline{T}$ of topological spaces is a closed simplicial model category.

Let $Sing:\underline{T} \to \underline{S}$ be the singular complex functor so that

$$(1) \qquad Hom_{\underline{S}}(K,Sing\ X) = Hom_{\underline{T}}(|K|,X)$$

(Actually Sing and $|\ |$ are adjoint simplicial functors which means that Hom can be replaced by $\underline{Hom}$ in (1).)

Lemma 1: The following are equivalent for a map f in $\underline{T}$.

(i) f is a fibration.

(ii) Sing f is a fibration in $\underline{S}$.

(iii) f has the RLP with respect to $|V(n,k)| \hookrightarrow |\Delta(n)|$ for $0 \leq k \leq n > 0$.

Proof: (ii) and (iii) are equivalent by (1), and (i) and (iii) are equivalent since $|V(n,k)| \hookrightarrow |\Delta(n)|$ is isomorphic in $\underline{T}$ to $I^{n-1} \times 0 \hookrightarrow I^n$.

Lemma 2: The following are equivalent for a map f in $\underline{T}$.

(i) f is a trivial fibration.

(ii) Sing f is a trivial fibration in $\underline{S}$.

(iii) f has the RLP with respect to $|\dot{\Delta}(n)| \hookrightarrow |\Delta(n)|$ for $n \geq 0$.

Proof: (ii) and (iii) are equivalent by (1). As $|\dot{\Delta}(n)| \hookrightarrow |\Delta(n)|$ is isomorphic in $\underline{T}$ to $S^{n-1} \subset D^n$ (where $S^{-1} = \emptyset$ if $n = 0$), the equivalence of (i) and (iii) becomes a standard obstruction theory argument which we omit.

Corollary: In $\underline{T}$ every object is fibrant and the fibrations and trivial fibrations satisfy SM7(a).

Proof: Since $\mathrm{Sing}(X^{|K|}) = (\mathrm{Sing}\ X)^K$ SM7(a) for $\underline{S}$ implies SM7(a) for $\underline{T}$.

Lemma 3: Any map f may be factored $f = pi$ where i is

a cofibration and p is a trivial fibration.

<u>Proof</u>: Letting $f:X \to Y$ we construct a diagram

$$\begin{array}{ccccccc} X & \xrightarrow{j_0} & Z^0 & \xrightarrow{j_1} & Z^1 & \longrightarrow & \cdots \\ & {\scriptstyle f}\searrow & \downarrow{\scriptstyle p_0} & \swarrow{\scriptstyle p_1} & & & \\ & & Y & & & & \end{array}$$

as follows. Let $Z^{-1} = X$ and $p_{-1} = f$, and having obtained Z^{n-1}, consider the set $\underline{D}$ of all diagrams D of the form

$$\begin{array}{ccc} |\dot{\Delta}(q_D)| & \xrightarrow{\alpha_D} & Z^{n-1} \\ \bigcap\downarrow & & \downarrow{\scriptstyle p_{n-1}} \\ |\Delta(q_D)| & \xrightarrow{\beta_D} & Y \end{array} \qquad q_D \geq 0$$

and define $j_n: Z^{n-1} \to Z^n$ by a co-cartesian diagram

$$\begin{array}{ccc} \bigvee_{D \in \underline{D}} |\dot{\Delta}(q_D)| & \hookrightarrow & \bigvee_{D \in \underline{D}} |\Delta(q_D)| \\ \downarrow{\scriptstyle \Sigma\alpha_D} & & \downarrow{\scriptstyle in_2} \\ Z^{n-1} & \xrightarrow{j_n = in_1} & Z^n \end{array}$$

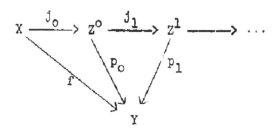

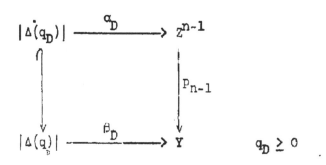

Define $p_n: Z^n \to Y$ by $p_n j_n = p_{n-1}$, $p_n in_2 = \Sigma\beta_D$, let $Z = \varinjlim Z^n$, $p = \varinjlim p_n$ and $i = \varinjlim j_n \circ \cdots \circ j_0$. By Lemma 1 j_n has the LLP with respect to trivial fibrations, hence i does too and so i is a cofibration. Now as $|\dot{\Delta}(n)|$ is compact any map $\alpha: |\dot{\Delta}(n)| \to Z$ factors through Z^m for m sufficiently large. In effect the well-known argument works because all the points of $Z - i(X)$ are closed. Hence given $\alpha: |\dot{\Delta}(n)| \to Z$, $\beta: |\Delta(n)| \to Y$ with $p\alpha$ = the restriction of β, there is an m with $\operatorname{Im}\alpha \subset Z^m$, and hence by the construction of Z^{m+1} a map $\gamma: |\Delta(n)| \to Z^{m+1} \subset Z$ such that $p\alpha = \beta$ and α = the restriction of γ to $|\dot{\Delta}(n)|$. By Lemma 1, p is a trivial fibration. Q.E.D.

<u>Remark</u>: The argument used to prove Lemma 3 relied primarily on the fact that $\operatorname{Hom}(|\dot{\Delta}(n)|, \varinjlim Z^m) = \varinjlim_{m} \operatorname{Hom}(|\dot{\Delta}(n)|, Z^m)$ and may be used to prove factorization whenever the fibrations (or trivial fibrations) are characterized by the RLP with respect to a set of maps $\{A_i \to B_i\}$ where each A_i is "sequentially small" in the sense that $\operatorname{Hom}(A_i, \cdot)$ commutes with sequential inductive limits. We will have further occasions to use this argument and will refer to it as the <u>small object argument</u>.

<u>Lemma 4</u>: The following are equivalent for a map $i: A \to B$.

(i) i is a trivial cofibration

(ii) i has the LLP with respect to the fibrations

(iii) i is a cofibration and a strong deformation retract map.

<u>Proof</u>: (iii) $\Longrightarrow$ (i) since a strong deformation retract map is a homotopy equivalence and hence a weak homotopy equivalence.

(ii) $\Longrightarrow$ (iii). Any trivial fibration is a fibration so i is a cofibration. The retract and strong deformation may be constructed by lifting in

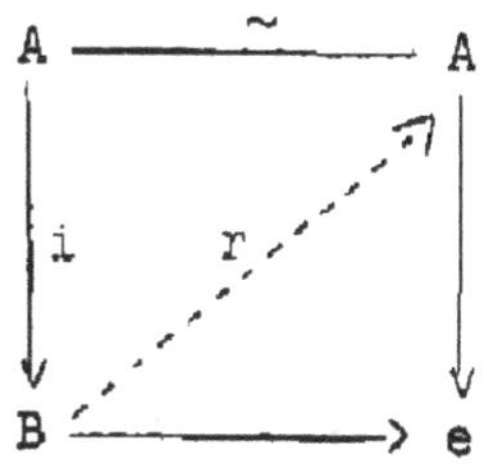

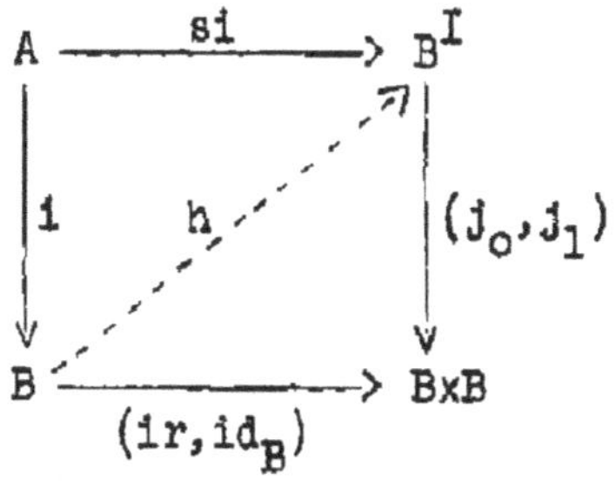

which is possible since $A \to e$ and $B^I \to B{\times}B$ are fibrations by the corollary to Lemma 2.

(iii) $\Longrightarrow$ (ii). A lifting in the first diagram, where p is a fibration, may be constructed by lifting in the second

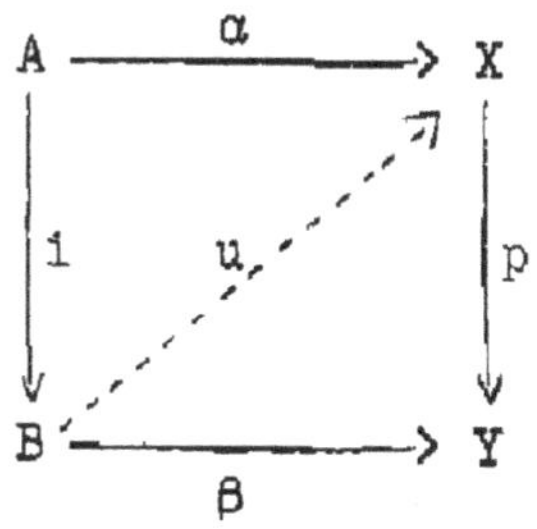

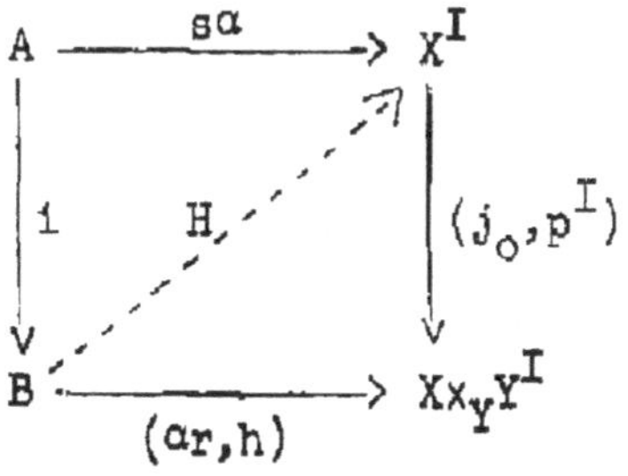

and setting $u = j_1 H$. Here r and h are the retract and strong deformation for i and lifting in the second diagram is possible because (j_0, p^I) is a trivial fibration by the corollary of Lemma 2.

(i) $\Longrightarrow$ (iii). Consider the following factorization of i

$$\begin{array}{ccc} A & \xrightarrow{j} & A \times_B B^I \\ & \searrow^{i} & \downarrow p \\ & & B \end{array}$$

which is the dual of the mapping cylinder construction. j is a strong deformation retract map, hence a weak equivalence, and p is a fibration. But i is a weak equivalence and so p is a trivial fibration. As i is a cofibration there is a section u of p with $ui = j$. Hence i is a retract of j and so i is a strong deformation retract map. Q.E.D.

<u>Proof of Theorem 1</u>: Axioms M0, SM0, and M5 are clear. Axiom M6 follows immediately from definitions and lemmas 1, 2 and 4. M6 and the corollary to Lemma 2 yield SM7. Lemma 3 gives one case of M2; to obtain the other, take $f: X \to Y$ and factor it $X \xrightarrow{j} X \times_Y Y^I \xrightarrow{p} Y$ where p is a fibration and j is a weak equivalence. Then factor $j = qi$ by Lemma 3 where i is a cofibration and q is a trivial fibration. By M5 i is a trivial cofibration hence $f = (qp)i$ is the desired factorization. This proves M2 and hence the theorem. Q.E.D.

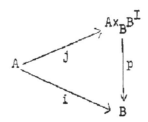

Let $\underline{G}$ be the category of simplicial groups endowed with its natural simplicial structure (see §1). Then $G \otimes K$ and G^K exist if $G \in \mathrm{Ob}\underline{G}$ and K is a simplicial set. In fact $(G \otimes K)_q = \bigvee_{\sigma \in K_q} G_q$ with natural simplicial operations and G^K is the function complex $\underline{\mathrm{Hom}}_{\underline{S}}(K,G)$ with its natural group structure. Define the normalization of $\underline{G}$ by

$$N_q(G) = \bigcap_{i>0} \mathrm{Ker}(d_i : G_q \to G_{q-1}) \qquad (= G_o \text{ if } q = 0)$$

$$d : N_q(G) \to N_{q-1}(G) \quad \text{induced by } d_o . \quad (= 0 \text{ if } q = 0)$$

and the (Moore) homotopy groups of G by

$$\pi_q(G) = \frac{\mathrm{Ker}\ d : N_q G \to N_{q-1} G}{\mathrm{Im}\ d : N_{q+1} G \to N_q G}$$

Then $\pi_q(G)$ is abelian for $q \geq 1$ and $\pi_o(G)$ is the set of components of G as a simplicial set.

A map in $\underline{G}$ will be called a weak equivalence if it induces isomorphisms for the functor π_*. A map will be called a fibration if it is a fibration as a map of simplicial sets and a cofibration if it has the LLP with respect to the trivial fibrations

<u>Theorem 2</u>: With these definitions the category $\underline{G}$ of simplicial groups is a closed simplicial model category.

The proof will be exactly the same as for topological spaces once we get the corollary of Lemma 2 for $\underline{G}$ and the homotopy axiom for the functor π_* .

Proposition 1: The following are equivalent for a map $f:G \to H$ of simplicial groups

(i) f is a fibration in $\underline{S}$ (hence in $\underline{G}$).

(ii) $N_q f: N_q G \to N_q H$ is surjective for $q > 0$.

(iii) $G \xrightarrow{(f,\varepsilon)} H \times_{K(\pi_0 H,0)} K(\pi_0 G,0)$ is surjective (in each dimension).

Here if A is a group we let $K(A,0)$ be the constant simplicial group which is A in each degree and which has all $\varphi^* = id_A$. It is readily verified that $G \mapsto \pi_0(G)$ is adjoint to $A \mapsto K(A,0)$, that is

$$\mathrm{Hom}_{\underline{G}}(G,K(A,0)) = \mathrm{Hom}_{(\mathrm{gps.})}(\pi_0(G),A)$$

and $\varepsilon: G \to K(\pi_0 G,0)$ is the adjunction map. The above proposition is essentially an elaboration of the following well-known fact which we shall assume.

Corollary: (Moore). A simplicial group is a Kan complex.

We shall also need the following fact which may be proved in exactly the same way as (Dold-Puppe 3.17).

Lemma 5: $f:G \to H$ is surjective (resp. injective) iff

$Nf: NG \to NH$ is surjective (resp. injective).

Proof of Proposition 1: (i) $\Longrightarrow$ (ii) since (ii) is equivalent to lifting in any diagram of the form

$$\begin{array}{ccc} V(n,0) & \xrightarrow{\;0\;} & G \\ \downarrow & & \downarrow f \\ \Delta(n) & \longrightarrow & H \end{array}$$

where 0 denotes the map sending all simplices to the identity elements of $\underline{G}$

(ii) $\Longrightarrow$ (iii). By Lemma 5 it suffices to show that $N(f,\varepsilon)$ is surjective. As N is left exact and $N_jK(A,0) = \{1\}$ $j > 0$ and A if $j = 0$, we find that $N_j(H \times_{K(\pi_0 H,0)} K(\pi_0 G,0)) = N_j H$ for $j > 0$, and hence $N_j(f,\varepsilon)$ is surjective for $j > 0$. It remains to show that $G_0 \to H_0 \times_{\pi_0 H} \pi_0 G$ is surjective which follows immediately by diagram chasing in the diagram

$$\begin{array}{ccccccc} N_1G & \xrightarrow{\;d\;} & G_0 & \longrightarrow & \pi_0 G & \longrightarrow & 1 \\ \downarrow & & \downarrow & & \downarrow & & \\ N_1H & \xrightarrow{\;d\;} & H_0 & \longrightarrow & \pi_0 H & \longrightarrow & 1 \end{array} .$$

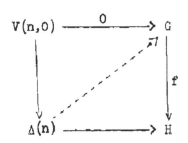

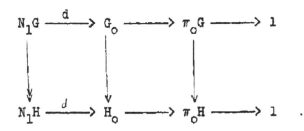

(iii) $\Longrightarrow$ (i). First suppose $f: G \to H$ is surjective. Given $u: V(n,k) \to G$ covering $v: \Delta(n) \to H$ we may extend u to a map $u': \Delta(n) \to G$ by the corollary. We may solve the lifting problem for u and v iff we may solve it for $0: V(n,k) \to G$ and $v \cdot (fu')^{-1}: \Delta(n) \to H$. Hence we reduce to the case $u = 0$. As f is surjective there is a map $w: \Delta(n) \to G$ with $fw = v$. Then $w|V(n,k)$ maps $V(n,k)$ to $\operatorname{Ker} f$ and by the corollary there is a $z: \Delta(n) \to \operatorname{Ker} f$ with $z|V(n,k) = w|V(n,k)$. Then $w \cdot z^{-1}: \Delta(n) \to G$ satisfies $(w \cdot z^{-1})|V(n,k) = 0 = u$ and $f \circ (w \cdot z^{-1}) = f \circ w = v$, thus providing the desired lifting. Hence any surjective map of simplicial groups is a fibration.

Returning to the general case we consider the diagram

$$\begin{array}{ccccc} G & \xrightarrow{(f,\epsilon)} & H \times_{K(\pi_0 H,0)} K(\pi_0 G,0) & \xrightarrow{pr_1} & H \\ & & \downarrow & & \downarrow \\ & & K(\pi_0 G,0) & \xrightarrow{K(\pi_0 f,0)} & K(\pi_0 H,0) \end{array}$$

where the square is cartesian. $K(\pi_0 f,0)$ is clearly a fibration hence so is pr_1, and (f,ϵ) being surjective is a fibration. Hence $f = pr_1(f,\epsilon)$ is a fibration. Q.E.D.

<u>Corollary</u>: f is surjective iff f is a fibration and $\pi_0(f)$ is surjective.

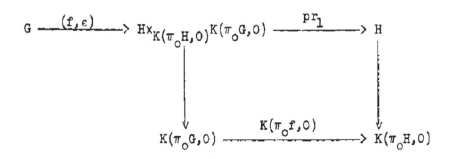

Proposition 2: The following are equivalent for a map f in $\underline{G}$.

(i) f is a trivial fibration in $\underline{S}$.

(ii) f is a trivial fibration in $\underline{G}$.

(iii) f is surjective and $\pi_*(\mathrm{Ker}\ f) = 0$.

Proof: (ii) $\Longleftrightarrow$ (iii). First of all the above corollary shows that f is surjective in case (ii). Letting K be the kernel of f we have the exact sequence of non-abelian group complexes

$$1 \to N(K) \to N(G) \to N(H) \to 1$$

where exactness at $N(K)$ at $N(G)$ is because N is left exact and exactness at $N(H)$ comes from Lemma 5. From this one gets by the usual diagram chasing a long exact sequence

$$\to \pi_1(G) \to \pi_1(H) \to \pi_o(K) \to \pi_o(G) \to \pi_o(H) \to 1$$

which show that $\pi_*(K) = 0$ iff $\pi_*(f)$ is an isomorphism.

(i) $\Longrightarrow$ (iii). First of all a trivial fibration is surjective in dimension 0 since it has the RLP with respect to $\dot{\Delta}(0) \subset \Delta(0)$; hence by Corollary 2 f is surjective. Next if $\alpha \in \pi_q(\mathrm{Ker}\ f)$ we represent α by $x \in K_q$ with $d_j x = 0 \quad 0 \leq j \leq q$ and define $u : \dot{\Delta}(q+1) \to \mathrm{Ker}\ f$ by sending all faces to the identity element of $\mathrm{Ker}\ f$ except the 0-th which goes to x. Lifting in

$$\begin{array}{ccc} \dot{\Delta}(q+1) & \xrightarrow{\ u\ } & G \\ \big\downarrow & & \big\downarrow f \\ \Delta(q+1) & \xrightarrow{\ 0\ } & H \end{array}$$

we obtain $y \in N_{q+1}(\text{Ker } f)$ with $dy = x$ showing that $\alpha = 0$.

(ii) + (iii) $\Longrightarrow$ (i). Given $u:\dot{\Delta}(n) \to G$ covering $v:\Delta(n) \to H$ we may lift if $n = 0$ since f is surjective. If $n > 0$, then as f is a fibration we may find $w:\Delta(n) \to G$ with $w|V(n,0) = u|V(n,0)$ and $fw = u$. Lifting for u and v is equivalent to lifting for $u \cdot w^{-1}$ and 0 so we reduce to the case $v = 0$ and $u|V(n,0) = 0$. Then u applied to the 0-th face of $\dot{\Delta}(n)$ is an element x of $(\text{Ker } f)_{n-1}$ with all faces the identity element. As $\pi_*(\text{Ker } f) = 0$, there is a $z \in N_n(\text{Ker } f)$ with $dz = x$. Then $\tilde{z}:\Delta(n) \to G$ satisfies $\tilde{z}|\dot{\Delta}(n) = u$ and $f\tilde{z} = 0$; hence $\tilde{z}$ is the desired lifting. Q.E.D.

Corollary: Every object of $\underline{G}$ is fibrant and the fibrations and trivial fibrations of $\underline{G}$ satisfy SM7(a).

Lemma 6: If $f,g:G \rightrightarrows H$ are homotopic maps in $\underline{G}$, then $\pi_*(f) = \pi_*(g):\pi_*(G) \to \pi_*(H)$.

Proof: We may assume that f is strictly homotopic to g.

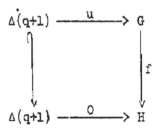

Let $h: G\times\Delta(1) \to H$ be a homotopy with $hi_0 = f$, $hi_1 = g$. Then $h = \{h_\sigma\}$ where σ is a simplex of $\Delta(1)$, $h_\sigma: G_q \to H_q$ is a group homomorphism and q is the degree of σ. σ may be identified with the sequence $(\sigma 0,\ldots,\sigma q)$ which is a sequence $(0\ldots 0,1\ldots 1)$. Let $h_i: G_q \to H_q$ be h_σ where σ has $i+1$ zeroes and $q-1$ ones. Then $h_{-1} = f$ and $h_q = g$ in degree q. If $\alpha \in \pi_q G$, represent α by $x \in G_q$ with $d_j x = 1$ $0 \le j \le q$, and set

$$z_1 = (h_0 s_0 x)\cdot(h_1 s_1 x)^{-1} \ldots (h_q s_q x)^{(-1)^q}$$

$$z_2 = (f s_0 x)\cdot(f s_1 x)^{-1} \ldots (f s_q x)^{(-1)^q}$$

Then $z_1 z_2^{-1} \in N_{q+1}H$ and $d(z_1 z_2^{-1}) = gx\cdot(fx)^{-1}$ showing that $\pi_q(f)\alpha = \pi_q(g)\alpha$. Q.E.D.

Proof of Theorem 2: We first note that Lemma 4 holds in $\underline{G}$. In effect (iii) $\Longrightarrow$ (i) because a homotopy equivalence is a weak equivalence by Lemma 6 and the rest of the proof used only the definition of cofibration and the corollary to Lemma 2 which for $\underline{G}$ is replaced by the corollary to Prop. 2. The factorization axiom Lemma 2 may be proved by the small object argument since trivial fibrations are characterized by the RLP with respect to $F\dot{\Delta}(n) \to F\Delta(n)$ (F = free group functor), and since $F\dot{\Delta}(n)$ is small. The rest of the proof follows that of Theorem 1. Q.E.D.

Let the category $\underline{S}$ of simplicial sets be considered as a simplicial category as in §1. Define fibrations and trivial fibrations as in §2 and call a map a cofibration (resp. trivial cofibration) if it has the LLP with respect to the class of trivial fibrations (resp. fibrations). Finally define a weak equivalence in $\underline{S}$ to be a map f which may be factored $f = pi$ where i is a trivial cofibration and p is a trivial fibration.

<u>Theorem 3</u>: With these definitions the category $\underline{S}$ of simplicial sets is a closed simplicial model category.

<u>Proof</u>: First note that "trivial" has its customary meaning in the sense that a map is a trivial cofibration (resp. fibration) iff it is a cofibration (resp. fibration) and a weak equivalence. Indeed the direction ($\Longrightarrow$) is clear. If $f:A \to B$ is a cofibration and

(2)

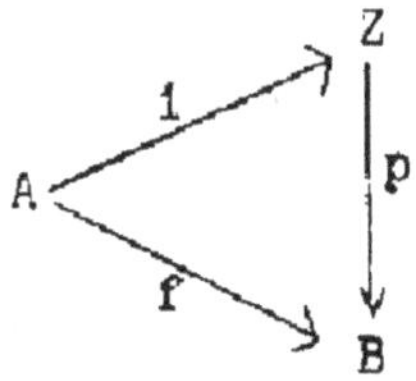

is a factorization of f, where i is a trivial cofibration and p is a trivial fibration, then there exists a section s of p with $sf = i$. Hence f is a retract of i and so f is a trivial cofibration. Fibrations are handled similarly.

The factorization axiom M2 may be proved by the small object argument using Prop. 1(i) and 2(i) of §2 and the fact that $\dot{\Delta}(n)$ and $V(n,k)$ are small. This actually proves that any map f

may be factored $f = pi$ where p is a trivial fibration (resp. fibration) and where i is a sequential composition of cobase extensions of direct sums of the maps $\dot{\Delta}(n) \to \Delta(n)$ (resp. $V(n,k) \to \Delta(n)$). In particular i is injective (resp. an "anodyne extension" in the terminology of Gabriel-Zisman). If f is already a cofibration (resp. trivial cofibration), then as above (see(2)) f is a retract of i, hence is injective (resp. an "anodyne extension"). The converse is also true (Prop. 1(iii), §1 and Gabriel-Zisman 3.1). Hence:

Proposition 2: In $\underline{S}$ the cofibrations are the injective maps and the trivial cofibrations are the anodyne extensions. Any object of $\underline{S}$ is cofibrant.

All of the axioms except M5 are now clear. M0, SM0 are trivial and M6 is true by the way things have been defined. M2 follows from the small object argument, and as the fibrations and trivial fibrations of $\underline{S}$ satisfies SM7(a), M6 implies that SM7 holds.

The fibrant objects of $\underline{S}$ are the Kan complexes. If E is a Kan complex and A is a simplicial set, then by SM7 $\underline{\text{Hom}}(A,E)$ is a Kan complex so "is strictly homotopic to" is an equivalence relation on $\text{Hom}(A,E)$. Let $[A,E] = \pi_0\underline{\text{Hom}}(A,E)$ denote the equivalence classes. Then M5 follows immediately from:

Proposition 3: A map $f:X \to Y$ in $\underline{S}$ is a weak equivalence if and only if for all Kan complexes E, $[f,E]:[Y,E] \to [X,E]$ is bijective.

<u>Proof</u>: ($\Longrightarrow$) If f is a trivial cofibration then this follows from the covering homotopy extension theorem (Prop. 4 §2) which depends only on SM7. If f is a trivial fibration then as every simplicial set is cofibrant one sees by the dual of the argument used to prove (ii)$\Longrightarrow$ (iii) in Lemma 3 that f is the dual of a strong deformation retract map. In particular f is a homotopy equivalence so $[f,E]$ is bijective. If f is a weak equivalence then f is the composition of a trivial cofibration and a trivial fibration so $[f,E]$ is bijective.

($\Longleftarrow$). Factoring $f = pi$ where i is a cofibration and p is a trivial fibration we have $[p,E]$ bijective by the above and so we reduce to the case where f is a cofibration. In this case f is a trivial cofibration by the following two lemmas.

<u>Lemma 7</u>: If i is a cofibration and $[i,E]$ is bijective for all Kan complexes E, then i has the LLP with respect to all fibrations of Kan complexes.

<u>Lemma 8</u>: If a cofibration i has the LLP with respect to all fibrations of Kan complexes, then it has the LLP with respect to all fibrations and so is a trivial cofibration.

<u>Proof of Lemma 7</u>: We begin by showing that if $p:X \to Y$ is a fibration of Kan complexes, then p is a trivial fibration if and only if p is a homotopy equivalence. The direction $\Longrightarrow$ has been proved above. To prove $\Longleftarrow$ let s be a homotopy inverse

for p. By lifting the homotopy from ps to id_Y we may assume that $ps = id_Y$. Then id_X and sp are homotopic and as X is a Kan complex we may choose $h: X \times \Delta(1) \to X$ with $hi_0 = sp$ and $hi_1 = id_X$. Now $\underline{Hom}(X,p): \underline{Hom}(X,X) \to \underline{Hom}(X,Y)$ is a fibration and the 1-simplices h and sph define a map $\alpha: V(2,0) \to \underline{Hom}(X,X)$ which covers the map $\beta: \Delta(2) \to \underline{Hom}(X,X)$ given by the 2-simplex $s_1(ph)$. Hence there is a map $\gamma: \Delta(2) \to \underline{Hom}(X,X)$ which covers β and restricts to α; the 0-th face of $\gamma(id)$ is a homotopy $k: X \times \Delta(1) \to X$ from id_X to sp which is fiber-wise, i.e. $pk = \sigma(p \times \Delta(1))$. This shows that $p: X \to Y$ is a fibration and the dual of a strong deformation retract and hence is a trivial fibration.

Now let $i: A \to B$ and E be as in the statement of Lemma 7 and apply this fact to the fibration $\underline{Hom}(i,E): \underline{Hom}(B,E) \to \underline{Hom}(A,E)$. If K is any simplicial set, then $[K, \underline{Hom}(B,E)] \to [K, \underline{Hom}(A,E)]$ may be identified with $[B, \underline{Hom}(K,E)] \to [A, \underline{Hom}(K,E)]$ which is bijective since $\underline{Hom}(K,E)$ is a Kan complex and the assumption on i. Hence $\underline{Hom}(i,E)$ is a trivial fibration.

Let $p: X \to Y$ be a fibration in $\underline{S}$ where Y and hence X is a Kan complex and consider the diagram

$$\begin{array}{ccccc} \underline{Hom}(B,X) & \xrightarrow{(i^*,p_*)} & \underline{Hom}(A,X) \times_{\underline{Hom}(A,Y)} \underline{Hom}(B,Y) & \xrightarrow{pr_1} & \underline{Hom}(A,X) \\ & & \downarrow{\scriptstyle pr_1} & & \downarrow{\scriptstyle \underline{Hom}(A,p)} \\ & & \underline{Hom}(B,Y) & \xrightarrow{\underline{Hom}(i,Y)} & \underline{Hom}(A,Y) \end{array}$$

where the square is cartesian. We have just shown that $\underline{\mathrm{Hom}}(i,Y)$ is a trivial fibration and hence so is pr_1 . Thus pr_1 and $pr_1(i^*,p_*) = i^* = \underline{\mathrm{Hom}}(i,X)$ are trivial fibrations, hence homotopy equivalences, and so (i^*,p_*) is a homotopy equivalence. As (i^*,p_*) is a fibration (SM7) it is a trivial fibration hence surjective in dimension zero and so i has the LLP with respect to p . Q.E.D.

Proof of Lemma 8: If $p:X \to Y$ is an arbitrary fibration in $\underline{S}$, then by (Barratt-Guggenheim-Moore)[4] there is a minimal fibration $q:Z \to Y$ such that Z is a strong deformation retract of X over Y (i.e. the homotopies are fiber-wise). As i is a cofibration SM7 implies that i has the LLP with respect to p iff i has the LLP with respect to q . But q is induced from a fibration of Kan complexes. To see this we may suppose Y is connected and let F be the fiber of q over a 0-simplex of Y . Then by (Barratt-Guggenheim-Moore)[4] there is a cartesian square

$$\begin{array}{ccc}
X & \longrightarrow & W(\underline{\mathrm{Aut}}\, F) \times_{\underline{\mathrm{Aut}}\, F} F \\
\Big\downarrow q & & \Big\downarrow r \\
Y & \longrightarrow & \overline{W}(\underline{\mathrm{Aut}}\, F)
\end{array}$$

where r is a fibration and $\overline{W}(\underline{\mathrm{Aut}}\, F)$ is a Kan complex. As i

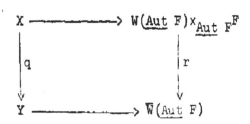

has the LLP with respect to r it does so also for q , and hence i is a trivial cofibration. This completes the proof of Lemma 8 and hence also of Theorem 3. Q.E.D.

Combining Prop. 2 with Prop. 1, §5, Ch.I, we obtain

Corollary: The anodyne extensions are precisely the injective maps in $\underline{S}$ which become isomorphisms in the homotopy category.

Remark: We have presented what we consider to be the most elementary proof of Theorem 3. The problem is to characterize the weak equivalences in some way so that M5 becomes clear. We now present a list of different characterizations of the weak equivalences. Some of these may be used to give alternative proofs of M5 and will be useful later.

Proposition 4: The following assertions are equivalent for a map $f:X \to Y$ of simplicial sets:

(i) f is a weak equivalence (isomorphism in homotopy category).

(ii) $[Y,E] \simeq [X,E]$ for all Kan complexes E .

(iii) $|X| \to |Y|$ is a homotopy equivalence in $\underline{T}$.

(iv) $Ex^{\infty}X \to Ex^{\infty}Y$ is a homotopy equivalence in $\underline{S}$.

(v) $H^0(Y,S) \simeq H^0(X,S)$ for any set S , $H^1(Y,G) \simeq H^1(X,G)$ for any group G , and $H^q(Y,\underline{L}) \simeq H^q(X,f^*\underline{L})$ for any local coefficient system $\underline{L}$ of abelian groups on Y and $q \geq 0$.

(vi) $\pi_0 X \cong \pi_0 Y$, $\pi_1(X,x) \cong \pi_1(Y,fx)$ for any $x \in X_0$, and $H^q(Y,L) \cong H^q(X,f^*\underline{L})$ where $\underline{L},q$ are as in (v).

<u>Proof</u>: (i) $\Longleftrightarrow$ (ii) is Proposition 5. (ii) $\Longleftrightarrow$ (iii) $\Longleftrightarrow$ (i are proved in [12]. Here $X \to Ex^{\infty}X$ is the functorial embedding of X into a Kan complex constructed by Kan.

(v) $\Longleftrightarrow$ (vi). Here $H^0(X,S) = Hom(\pi_0 X,S)$, $H^1(X,G) = [X,\overline{W}(G)]$ and $\pi_1(X,x)$ is the fundamental group of X at x calculated by the method of the maximal tree. The first assertion of (v) and (vi) are equivalent and we may assume X and Y are connected. Let $x \in X_0$. Then $[X,\overline{W}(G)] = Hom_{(gps.)}(\pi_1(X,x),G)_G$ where G acts on a homomorphism φ by $(g\cdot\varphi)(\lambda) = g\varphi(\lambda)g^{-1}$. In other words $[X,\overline{W}(G)]$ is the set of homomorphisms from $\pi_1(X,x)$ to G in the category of groups up to inner automorphisms, so the second condition of (v) means that $\pi_1(X,x) \to \pi_1(Y,fx)$ is an isomorphism in this category. But this is clearly the same as $\pi_1(X,x) \to \pi_1(Y,fx)$ being an isomorphism of groups, and so we see that the second conditions of (v) and (vi) are equivalent. Thus (v) and (vi) are equivalent.

(iii) $\Longrightarrow$ (vi). As $\pi_0|X| = \pi_0 X$ we may assume X and Y are connected. As $\pi_1(|X|,x) = \pi_1(X,x)$ we conclude that $\pi_1(X,x) \cong \pi_1(Y,fx)$ for all $x \in X_0$. Let x_0 be a fixed 0-simplex of X , let $y_0 = fx_0$ and let $\pi = \pi_1(X,x_0) \simeq \pi_1(Y,y_0)$. Let $p:(\tilde{X},\tilde{x}_0) \to (X,x)$ (resp. $q:(\tilde{Y},\tilde{y}_0) \to (Y,y_0)$) be the universal coverings and $\tilde{f}:\tilde{X} \to \tilde{Y}$ the unique map covering f with $\tilde{f}\tilde{x}_0 = \tilde{y}_0$.

If $\underline{L}$ is a local coefficient system on Y, then there is a morphism of Cartan-Leray spectral sequences

$$\begin{array}{ccc} E_2^{pq} = H^p(\pi, H^q(\tilde{Y}, q^*\underline{L})) & \Longrightarrow & H^{p+q}(Y,\underline{L}) \\ \downarrow & & \downarrow \\ E_2^{pq} = H^p(\pi, H^q(\tilde{X}, p^*f^*\underline{L})) & \Longrightarrow & H^{p+q}(X, f^*\underline{L}) \end{array}$$

As $|\tilde{X}|$ and $|\tilde{Y}|$ are the universal coverings of X and Y, (iii) $\Longrightarrow$ $|\tilde{f}|$ is a homotopy equivalence. As $H^*(|\tilde{X}|,A) = H^*(\tilde{X},A)$ for any abelian group A we see that the map on the E_2 is an isomorphism and so (vi) is proved.

(vi) $\Longrightarrow$ (iii). We may assume X and Y are connected and we let $\tilde{X}, \tilde{Y}, \pi$, etc., be as above. By a theorem of Whitehead it suffices to prove that $\pi_q(|X|,x_o) \simeq \pi_q(|Y|,y_o)$ for all q. For $q = 1$, this comes from $\pi_1(|X|,x_o) = \pi_1(X,x_o)$ and the similar assertion for Y. For $q > 1$ it suffices to prove $|\tilde{f}|$ is a homotopy equivalence or equivalently, since $|\tilde{X}|$ and $|\tilde{Y}|$ are 1-connected, that $H^*(\tilde{Y},A) \simeq H^*(\tilde{X},A)$ for any abelian group A. But the Leray spectral sequences for p and q degenerate giving a diagram

$$\begin{array}{ccc} H^n(Y,q_*A) & \simeq & H^n(\tilde{Y},A) \\ \downarrow{\scriptstyle f^*} & & \downarrow{\scriptstyle (\tilde{f})^*} \\ H^n(X,p_*A) & \simeq & H^n(\tilde{X},A) \end{array}$$

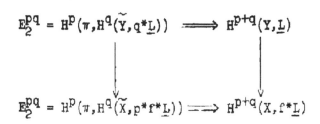

where p_*A, q_*A are the local coefficient systems of the cohomology of the fiber, and where f^* is the map on cohomology coming from $f^*(q_*A) = p_*A$. By (vi) f^* is an isomorphism and so we are finished. Q.E.D.

§4. <u>$s\underline{A}$ as a model category.</u>

Let $\underline{A}$ be a category closed under finite limits. A map $f:X \to Y$ is said to be an <u>effective epimorphism</u> if for any object T the diagram of sets

$$\mathrm{Hom}(Y,T) \xrightarrow{f^*} \mathrm{Hom}(X,T) \underset{pr_2^*}{\overset{pr_1^*}{\rightrightarrows}} \mathrm{Hom}(X\times_Y X,T)$$

is exact. We shall say that an object P of $\underline{A}$ is <u>projective</u> if $\mathrm{Hom}(P,X) \to \mathrm{Hom}(P,Y)$ is surjective whenever $X \to Y$ is an effective epimorphism and that $\underline{A}$ has <u>sufficiently many projectives</u> if for any object X there is a projective P and an effective epimorphism $P \to X$. If $\underline{A}$ is closed under inductive limits, we call an object X <u>small</u> if $\mathrm{Hom}(X,\cdot)$ commutes with <u>filtered</u> inductive limits, and call a class $\underline{U}$ of objects of $\underline{A}$ a class of <u>generators</u> if for every object X there is an effective epimorphism $Q \to X$, where Q is a direct sum of copies of members of $\underline{U}$.

<u>Theorem 4</u>: Let $\underline{A}$ be a category closed under finite limits and having sufficiently many projectives. Let $s\underline{A}$ be the simplicial category of simplicial objects over $\underline{A}$. Define a map f in $s\underline{A}$ to be a fibration (resp. weak equivalence) if $\underline{\mathrm{Hom}}(P,f)$ is a fibration (resp. weak equivalence) in $\underline{S}$ for each projective object P of $\underline{A}$, and a cofibration if f has the LLP with respect to the class of trivial fibrations. Then $s\underline{A}$ is a closed

simplicial model category if $\underline{A}$ satisfies one of the following extra conditions:

(*). Every object of $s\underline{A}$ is fibrant.

(**). $\underline{A}$ is closed under inductive limits and has a set ot small projective generators.

Here, and in the following, objects of $\underline{A}$ will be identified with constant simplicial objects. For the rest of this section $\underline{A}$ will denote a category closed under finite limits and having sufficiently many projectives. We will not use conditions (*) and (**) until we absolutely have to. We first make some remarks about the theorem.

Proposition 1: Suppose that every object X of $\underline{A}$ is a quotient of a cogroup object C (i.e. there exists an effective epimorphism $C \to X$). Then $\underline{A}$ satisfies (*).

Proof: Given $X \in \mathrm{Ob}\, s\underline{A}$ and a projective object P of $\underline{A}$, choose an effective epimorphism $C \to P$ where C is a cogroup object. Then P is a retract of C, so $\underline{\mathrm{Hom}}(P,X)$ is a retract of $\underline{\mathrm{Hom}}(C,X)$ which is a group complex. By Moore, $\underline{\mathrm{Hom}}(C,X)$ is a Kan complex hence so is $\underline{\mathrm{Hom}}(P,X)$, and hence X is fibrant.

Remarks: 1. By a theorem of Lawvere [4] a category closed under inductive limits and having a single small projective generator U is equivalent to the category of universal algebras with a specified set of finitary operations and identities in such a way that

U corresponds to the free algebra on one generator. Hence the theorem applies when $\underline{A}$ is the category of rings, monoids, etc. One may show that effective epimorphism = set-theoretically surjective map in this case.

2. The category of profinite groups satisfies (*) but not (**). The free profinite group generated by a profinite set is both projective and a cogroup object in this category and every object is a quotient of such an object.

The rest of this section contains the proof of Theorem 4.

Proposition 2: Let $\underline{A}$ be a category closed under finite limits and having sufficiently many projectives. Then $X \to Y$ is effective epimorphism $\Longleftrightarrow$ $\mathrm{Hom}(P,X) \to \mathrm{Hom}(P,Y)$ is surjective for every projective object P.

Proof: ($\Longrightarrow$) is by definition. For ($\Longleftarrow$) we first establish three properties of effective epimorphisms which hold without assuming $\underline{A}$ has enough projectives. It is clear that $f:X \to Y$ is an effective epimorphism iff for any object T and map $\alpha:X \to T$ there is a unique $\beta:Y \to T$ with $\beta f = \alpha$ provided α satisfies the necessary condition that $\alpha u = \alpha v$ whenever $u,v:S \rightrightarrows X$ are two maps such that $fu = fv$.

(1) If $f:X \to Y$ has a section $s:Y \to X$ with $fs = id_Y$, then f is an effective epimorphism.

In effect given $\alpha:X \to T$ satisfying the necessary condition let $\beta = \alpha s:Y \to T$. As sf, $id_X:X \rightrightarrows X$ are two maps with

$f(sf) = f(id_X)$ we have $\beta f = \alpha sf = \alpha$. β is clearly unique.

(2) If $X \xrightarrow{f} Y \xrightarrow{g} Z$ are maps, where gf is an effective epimorphism and f is an epimorphism, then g is an effective epimorphism.

Given $\alpha: Y \to T$ with $\alpha u = \alpha v$ whenever $u,v: S \to Y$ and $gu = gv$, it follows that $\alpha f: X \to T$ has the property that $\alpha fu = \alpha fv$, whenever $u,v: S \to Y$ and $gfu = gfv$. As gf is an effective epimorphism, there is a unique map $\beta: Z \to T$ with $\beta gf = \alpha f$. As f is an epimorphism $\beta g = \alpha$.

(3) If $X \xrightarrow{f} Y \xrightarrow{g} Z$ are maps, where g is an effective epimorphism and f has a section s , then gf is an effective epimorphism.

In effect given $\alpha: X \to T$ satisfying the necessary conditions that it factor through gf, it in particular satisfies the necessary conditions for factoring through f . By (1) there is a unique β with $\beta f = \alpha$ given by $\beta = \alpha s$. Suppose $u,v: S \rightrightarrows Y$ are such that $gu = gv$. Then $gfsu = gfsv$ so $\alpha su = \alpha sv$ or $\beta u = \beta v$. Hence since g is an effective epimorphism there is a unique γ with $\gamma g = \beta$ and hence a unique γ with $\gamma gf = \alpha$. Thus gf is an effective epimorphism.

Now suppose that $f: X \to Y$ has $\mathrm{Hom}(P,X) \to \mathrm{Hom}(P,Y)$ surjective for all projective objects P . Choose an effective epimorphism $u: P \to X$ with P projective. As fu has the same property as f we are reduced by (2) to the case where X is projective. Choose an effective epimorphism $v: Q \to Y$ with Q

projective. As X is projective there is a map $\alpha: X \to Q$ with $v\alpha = f$ and by the property of f there is a map $\beta: Q \to X$ with $f\beta = v$. The maps α and β yield sections of the maps pr_1 and pr_2 in

$$\begin{array}{ccc} X\times_Y Q & \xrightarrow{pr_2} & Q \\ \Big\downarrow{\scriptstyle pr_1} & & \Big\downarrow{\scriptstyle v} \\ X & \xrightarrow{f} & Y \end{array}$$

By (1) and (3) $vpr_2 = fpr_1$ is an effective epimorphism and so by (2) f is an effective epimorphism. Q.E.D.

<u>Corollary</u>: The class of effective epimorphisms in $\underline{A}$ is closed under composition and base change and it contains all isomorphisms. If gf is an effective epimorphism so is g.

In particular the effective epimorphisms are universally effective.

<u>Proposition 3</u>: Any map f may be factored $f = pi$ where i is a cofibration and where p is a trivial fibration.

<u>Proof</u>: Given $f: X \to Y$ construct a diagram

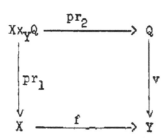

$$\begin{array}{ccccccc} X & \xrightarrow{j_o} & Z^o & \xrightarrow{j_1} & Z^1 & \xrightarrow{j_2} & \cdots \\ & \searrow^{f} & \downarrow^{p_o} & \swarrow_{p_1} & & & \\ & & Y & & & & \end{array}$$

as follows. Let $Z^{-1} = X$, $p^{-1} = f$ and having obtained $p_{n-1}: Z^{n-1}$ Y, choose a projective object P_n of $\underline{A}$ and a map (α,β) so tha

$$(1) \quad P_n \vee (Z^{n-1})^{\Delta(n)} \xrightarrow{(\alpha,\beta)+(p_{n-1}^{\Delta(n)},\,(Z^{n-1})^{i_n})} Y^{\Delta(n)} \times_{Y^{\dot{\Delta}(n)}} (Z^{n-1})^{\dot{\Delta}(n}$$

is an effective epimorphism in dimension 0, where $i_n: \dot{\Delta}(n) \to \Delta(n)$ is the canonical inclusion. Now define the map j_n by a co-cartesian diagram

$$(2) \qquad \begin{array}{ccc} P_n \otimes \dot{\Delta}(n) & \xrightarrow{P_n \otimes i_n} & P_n \otimes \Delta(n) \\ \downarrow \beta & & \downarrow in_2 \\ Z^{n-1} & \xrightarrow{in_1 = j_n} & Z^n \end{array}$$

and let $p_n = Z^n \to Y$ be the unique map with $p_n j_n = p_{n-1}$ and $p_n in_2 = \alpha$.

As $i_n: \dot{\Delta}(n) \to \Delta(n)$ is an isomorphism in dimensions $< n$ so is j_n, hence $\varinjlim Z^n = Z$ exists and we may define map $X \overset{i}{\to} Z \overset{p}{\to} Y$

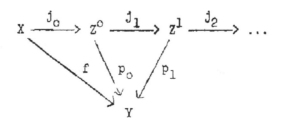

by $i = \lim j_n \cdots j_o$, $p = \lim p_n$. It is clear that $P_n \otimes i_n$ in (2) is a cofibration, hence each j_n and hence i is a cofibration. To see that p is a trivial fibration it suffices to show that $(p^{\Delta(n)}, z^{i_n}) : Z^{\Delta(n)} \to Y^{\Delta(n)} \times_{Y^{\dot{\Delta}(n)}} (Z^{n-1})^{\dot{\Delta}(n)}$ is an effective epimorphism in dimension 0. Consider the diagram

$$\begin{array}{ccc}
P_n \vee (Z^{n-1})^{\Delta(n)} & \xrightarrow{\qquad\qquad\qquad\qquad} & Y^{\Delta(n)} \times_{Y^{\dot{\Delta}(n)}} (Z^{n-1})^{\dot{\Delta}(n)} \\
\Big\downarrow {\scriptstyle \beta + j_n^{\Delta(n)}} & & \Big\downarrow {\scriptstyle (id, k_{n-1}^{\dot{\Delta}(n)})} \\
(Z^n)^{\Delta(n)} \xrightarrow{\ k_n^{\Delta(n)}\ } Z^{\Delta(n)} & \xrightarrow{(p^{\Delta(n)}, Z^{i_n})} & Y^{\Delta(n)} \times_{Y^{\dot{\Delta}(n)}} Z^{\dot{\Delta}(n)}
\end{array}$$

where the top map is the effective epimorphism (1), and where $k_q = \lim j_n \cdots j_{q+1}$. k_{q+1} is an isomorphism in dimensions $< n$, hence $(id, k_{n-1}^{\dot{\Delta}(n)})$ is an isomorphism in dimension zero. By the corollary of Prop. 2, $(p^{\Delta(n)}, z^{i_n})$ is an effective epimorphism in dimension 0. Q.E.D.

<u>Proof of Theorem 1</u>: (*). This is exactly the same as the proof in §3 for $\underline{T}$ and $\underline{G}$, so we present an outline only. If $f : A \to B$ is a map, then as A and B are fibrant, $A \xrightarrow{i} A \times_B B^I \xrightarrow{p} B$ is a factorization of f into a strong deformation retract map followed by a fibration. The homotopy equivalence i in $s\underline{A}$ is

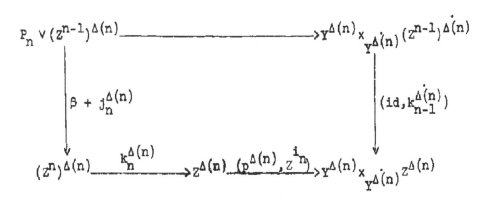

carried by $\underline{\mathrm{Hom}}(P,\cdot)$ into a homotopy equivalence in $\underline{S}$; hence i is a weak equivalence in $s\underline{A}$. If f has the LLP with respect to fibrations, f is a cofibration and a retract of i ; hence f is a trivial cofibration. Conversely if f is a trivial cofibratio M5 implies p is a trivial fibration so f is a retract of i ; hence f is a strong deformation retract map, so by SM7(a), f has the LLP with respect to the fibrations. With this we have M6, hence SM7. Finally M2 results from Prop. 3 for the cofibration--trivial fibration case and for the other case one uses this case to write $i = qj$, j cofibration, q trivial fibration, whence $f = (pq)j$ is a factorization where j is a trivial cofibration and pq is a fibration.

(**). Let $\underline{U}$ be a set of small projective generators for $\underline{A}$. Then the retract argument used in the proof of Prop. 1 shows that a map f in $s\underline{A}$ is a fibration or weak equivalence iff $\underline{\mathrm{Hom}}(P,f)$ is so is $\underline{S}$ for all $P \in \underline{U}$. In particular the fibrations are characterized by the RLP with respect to the set of maps $P \otimes V(n,k) \to P \otimes \Delta(n)$ for each $P \in \underline{U}$ and $0 \leq k \leq n > 0$. However $P \otimes V(n,k)$ is small in $s\underline{A}$ since P is small in $\underline{A}$, hence the small object argument implies that any map f may be factored $f = pi$ where p is a fibration and i has the LLP with respect to all fibrations. We must show that i is a weak equivalence.

For this purpose we shall use Kan's Ex^{∞} functor ([12]). We recall that $(\mathrm{Ex}\ K)_n = \mathrm{Hom}_{\underline{S}}(\mathrm{Sd}\Delta(n),K)$, where Sd is the subdivisio functor hence $(\mathrm{Ex}K)_n$ is the projective limit in the category of sets of a finite diagram involving K_n, K_{n-1} and the face operators of K. As $\underline{A}$ is closed under finite limits, we may define $Ex:s\underline{A} \to s\underline{A}$ by the formula

$$(3) \qquad \underline{Hom}(A, Ex\ X) = Ex\ \underline{Hom}(A,X)$$

for all $A \in Ob\,\underline{A}$, $X \in Ob\,s\underline{A}$. The natural map $K \to Ex\,K$ in $\underline{S}$ extends to a map $X \to Ex\,X$, and hence we may define $Ex^{\infty}(X) = \varinjlim Ex^n(X)$ and a map $\epsilon_X : X \to Ex^{\infty}(X)$. If $P \in \underline{U}$, then as P is small $\underline{Hom}(P, Ex^{\infty}X) = Ex^{\infty}\underline{Hom}(P,X)$. Therefore $Ex^{\infty}X$ is fibrant and $\epsilon_X : X \to Ex^{\infty}X$ is a weak equivalence.

Now suppose that $i:A \to B$ has the LLP with respect to fibrations. Then we may lift successively in

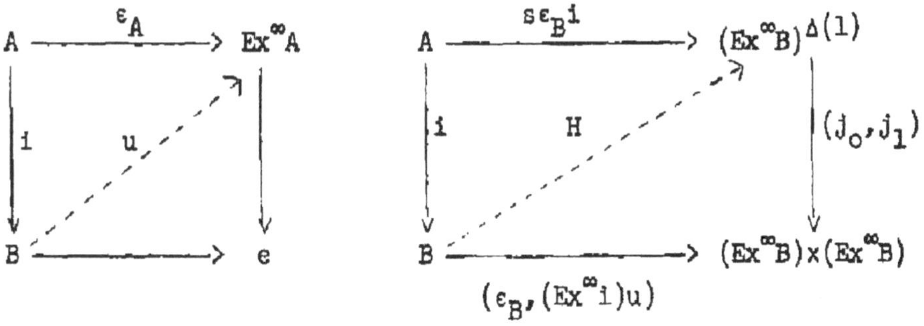

obtaining the formulas $ui = \epsilon_A$, $(Ex^{\infty}i)u \sim \epsilon_B$, $\epsilon_B i = (Ex^{\infty}i)\epsilon_A$. Let $P \in \underline{U}$ and apply the functor $\gamma \circ \underline{Hom}(P,\cdot)$ where γ is the canonical localization map $\underline{S} \to Ho\underline{S}$. It follows that $\gamma\underline{Hom}(P,i)$ is an isomorphism hence (Prop. 1, §5, Ch.I) $\underline{Hom}(P,i)$ is a weak equivalence. Thus i is a weak equivalence and we have proved that a map with the LLP with respect to the fibrations is a trivial cofibration. Conversely if f is a trivial cofibration we may factor $f = pi$ where p is a fibration and i has the LLP with respect to the fibrations; by what we have just shown i is a weak equivalence, hence p is trivial, so f is a retract of i and hence has the LLP with respect to the fibrations. This proves half of M6 and M2; the other is similar using Prop. 3. M6 implies SM7 and M5 is clear, so the theorem is proved. Q.E.D.

Remarks: 1. Some extra condition on $\underline{A}$ like (*) or (**) is necessary since the category of simplicial finite sets fails to satisfy M2. In effect there are simplicial finite sets with infinite homotopy groups.

2. If the map $\emptyset \to X$ in $s\underline{A}$ is factored $\emptyset \xrightarrow{i} Z \xrightarrow{p} X$ where p is a trivial fibration and i is a cofibration, then it is easily seen by using Prop. 4, §2 that this factorization is unique up to simplicial homotopy over X. Now for $Z \to X$ to be a trivial fibration is the analogue of Z being a resolution of X, while for Z to be cofibrant is the analogue of Z being a complex of projective objects. Hence Prop. 3 asserts for $s\underline{A}$ the existence of projective resolutions and so one may define derived functors for $\underline{A}$ even when $\underline{A}$ does not satisfy (*) or (**).

3. It is worthwhile noting that $(X^{\dot{\Delta}(n)})_o = (\text{cosk}_{n-1}X)_n$ where cosk_q is the q-th coskeleton functor of Verdier [19]. Consequently a trivial fibration $X \to A$ where A is an object of

$\underline{A}$ is the same as hypercovering of A for the Grothendieck topology whose covering families consist of single maps $\{v \to u\}$ which are effective epimorphisms. We will discuss this in the next section.

4. When $\underline{A}$ is a category of universal algebras (see Remark 1 after Prop. 1), then the P_n in the proof of Prop. 3 may be chosen to be free algebras, and so the map $X \xrightarrow{i} Z$ is free in the following sense: There are subsets $C_q \subset Z_q$ for each q such that

(i) $\eta^* C_p \subset C_q$ whenever $\eta:[q] \to [p]$ is a surjective monotone map,

(ii) $f_q + g_q : X_q \vee FC_q \to Z_q$ is an isomorphism for all q, where FC_q is the free algebra generated by C_q and $g_q : FC_q \to Z_q$ is the unique algebra map which is the identity on C_q. Conversely one may show ([13], th. 6.1) that any free map $X \xrightarrow{i} Z$ may be factored $X \to Z^0 \to Z^1 \to \ldots \to Z$ where there are co-cartesian squares (2) with P_n free and hence any free map is a cofibration. Furthermore given a cofibration f we may factor it $f = pi$ where i is free and p is a trivial fibration; then f is a retract of i hence <u>a map is a cofibration iff it is a retract of a free map</u>.

5. If $\underline{A}$ is an abelian category with sufficiently many projective objects, then Theorem 4 endows $s\underline{A}$ with the structure of a closed model category. On the other hand by Dold-Puppe [6] the normalization functor $N: s\underline{A} \to Ch\underline{A}$, the category of chain complexes in $\underline{A}$, is an equivalence of categories, and moreover the simplicial homotopy relation on maps in $s\underline{A}$ corresponds to the chain homotopy relation on maps in $Ch\underline{A}$. The corresponding closed model category

structure on Ch $\underline{A}$ may be described as follows: Weak equivalence are maps inducing isomorphisms on homology groups (since $H(NX) = \pi X$) and fibrations are maps which are epimorphisms in positive degrees (straightforward generalization of Prop. 1, §3 to abelian categories). Finally cofibrations are monomorphisms whose cokernels are dimension-wise projective. In effect what is called the fundamental theorem of homological algebra amounts essentially to the following: (i) any monomorphism with dimension-wise projective cokernel has the LLP with respect to trivial fibrations and (ii) any map f may be factored $f = pi$ where p is a trivial fibration and i is a monomorphism with dimension-wise projective cokernel.

As the class of monomorphisms with dimension-wise projective cokernels is closed under retracts, it is seen to be the class of cofibrations by a retract argument.

§5. Homology and cohomology.

If homotopical algebra is thought of as "non-linear" or "non-additive" homological algebra, then it is natural to ask what is the "linearization" or "abelianization" of this non-linear situation. This leads to a uniform description of homology and cohomology for model categories and in the case of $s\underline{A}$ the resulting cohomology agrees with the cohomology constructed using suitable cotriples and Grothendieck topologies.

Let $\underline{C}$ be a model category and let $\underline{C}_{ab}$ be the category of abelian group objects in $\underline{C}$. We assume that the abelianization X_{ab} of any object X of $\underline{C}$ exists so that there are adjoint functors

(1) $$\underline{C} \underset{i}{\overset{ab}{\rightleftarrows}} \underline{C}_{ab} .$$

where i is the faithful inclusion functor. We also assume that $\underline{C}_{ab}$ is a model category in such a way that these adjoint functors satisfy the conditions of the first part of Theorem 3, §4, Ch.I., so that there are adjoint functors

(2) $$\mathrm{Ho}\underline{C} \underset{\underline{\underline{R}}i}{\overset{\underline{\underline{L}}ab}{\rightleftarrows}} \mathrm{Ho}\,\underline{C}_{ab}$$
$$[X,\underline{\underline{R}}i(A)] = [\underline{\underline{L}}ab(X),A]$$

Finally we shall assume that $Ho\underline{C}_{ab}$ satisfies the following two conditions:

A. The adjunction map $\theta: A \cong \Omega\Sigma A$ is an isomorphism for all objects A.

B. If $A' \xrightarrow{i} A \xrightarrow{j} A'' \xrightarrow{\delta} \Sigma A'$ is a cofibration sequence, then $\Omega\Sigma A' \xrightarrow{-i\cdot\theta^{-1}} A \xrightarrow{j} A'' \xrightarrow{\delta} \Sigma A'$ is a fibration sequence (Note that as $Ho\underline{C}_{ab}$ is additive the action $F \times \Omega B \xrightarrow{m} F$ is determined by $\partial = m(0,id): \partial B \to F$ via the rule $m(\alpha,\lambda) = \alpha + \partial\lambda$ if $\alpha: T \to F$ and $\lambda: T \to \Omega B$).

These conditions hold for example if $\underline{C}_{ab} = s\underline{A}$, where $\underline{A}$ is an abelian category with enough projectives and if $\underline{C}_{ab}$ is the model category of simplicial modules over a simplicial ring (see following section).

We define the <u>cohomology</u> groups of an object X of $Ho\underline{C}$ with coefficients an object A of $Ho\underline{C}_{ab}$ to be

$$H_M^q(X,A) = [\underline{\underline{L}}\,ab(X), \Omega^{q+N}\Sigma^N A]$$

where N is an integer ≥ 0 with $q + N \geq 0$. By (A) it does not matter what N we choose. Suppose now that $\underline{C}$ is pointed. Then

$$H_M^q(\Sigma X,A) = [\underline{\underline{L}}\,ab(\Sigma X), \Omega^{q+N}\Sigma^N A] = [\Sigma\underline{\underline{L}}\,ab(X), \Omega^{q+N}\Sigma^N A]$$

$$= [\underline{\underline{L}}\,ab(X), \Omega^{q+N+1}\Sigma^N A] = H_M^{q+1}(X,A)$$

Using this and the fact that $\underline{\underline{L}}ab$ preserves cofibration sequences, we find that if $X \to Y \to C$, etc. is a cofibration sequence, then there is a long exact sequence

$$\to H^q_M(C,A) \to H^q_M(Y,A) \to H^q_M(X,A) \xrightarrow{\delta} H^{q+1}_M(C,A) \to$$

From (B) it follows that if $A' \to A \to A'' \to \Sigma A'$ is a cofibration sequence in $\mathrm{Ho}\underline{C}_{ab}$ then there is a long exact sequence

$$\to H^q_M(X,A') \to H^q_M(X,A) \to H^q_M(X,A'') \xrightarrow{\delta} H^{q+1}_M(X,A') \to$$

It is reasonable to call an object of $\mathrm{Ho}\underline{C}$ of the form $\underline{\underline{R}}i(A)$ a <u>generalized Eilenberg-Maclane object</u> and to call $\underline{\underline{L}}ab(X)$ the <u>homology</u> of X . In effect

$$H^o_M(X,A) = [\underline{\underline{L}}ab(X), A]$$

is a universal coefficient theorem while

$$H^o_M(X,A) = [X, \underline{\underline{R}}i(A)]$$

is a representability theorem for cohomology.

<u>Examples</u>: 1. $\underline{C} = \underline{S}$ so that $\underline{S}_{ab} = s(Ab)$ the category of simplicial abelian groups and $X_{ab} = \mathbb{Z}X$, the free abelian group

functor applied dimension-wise to X. The assumptions on $\underline{S}$ and $\underline{S}_{ab}$ hold and as every object of $\underline{S}$ is cofibrant $\underline{\underline{L}}ab(X) = X_{ab}$. Hence

$$H_M^*(X,K(R,0)) = H^*(X,R) ,$$

the usual cohomology of X with values in the abelian group R. Also

$$\pi_*(X_{ab}) = H_*(X,\mathbb{Z})$$

which partially justifies calling X_{ab} the homology of X.

2. Let $\underline{C} = \underline{G}$ so that $\underline{G}_{ab} = s(Ab)$ and $G_{ab} = G/[G,G]$. Then $\underline{\underline{L}}ab(G) = G_{ab}$ if G is a free simplicial group and so by a result of Kan [11] (see also §6 (16))

$$H_M^q(G,K(R,0)) = \begin{cases} 0 & q < 0 \\ H^{q+1}(\overline{W}G,R) & q \geq 0 \end{cases}$$

where $\overline{W}G$ is the "classifying space" simplicial set of G. Also

$$\pi_q(\underline{\underline{L}}ab(G)) = H_{q+1}(\overline{W}G,\mathbb{Z}) .$$

These formulas are seen to hold for any simplicial group G since to calculate $\underline{\underline{L}}ab(G)$ we may replace G by a free simplicial group.

We now show how these model cohomology groups compare with other kinds of cohomology. In the following $\underline{A}$ denotes a category closed under finite projective limits, X is an object of $\underline{A}$, and A is an abelian group object in $\underline{A}/X$. We consider four definitions of cohomology of X with values in A.

(1) Suppose that the effective epimorphisms of $\underline{A}$ are universal effective epimorphisms (which is the case if $\underline{A}$ has sufficiently many projectives--§4, Cor. to Prop. 2). We define a Grothendieck topology on $\underline{A}$ ([1]) by defining a covering of an object Y to be a family consisting of a single map $U \to Y$ which is an effective epimorphism. The induced topology on $\underline{A}/X$ is coarser than the canonical topology so the representable functor h_A is a sheaf of abelian groups; hence sheaf cohomology groups, which we shall denote by $H^*_{GT}(X,A)$, are defined. Thus $H^q_{GT}(X,A) = H^q\{I^{\cdot}(X)\}$ where $I^{\cdot}$ is an injective resolution of h_A in the category of abelian sheaves on X.

(2) Suppose that there are adjoint functors

$$\underline{A} \underset{S}{\overset{F}{\rightleftarrows}} \underline{B}$$

(3)
$$\mathrm{Hom}_{\underline{A}}(FB,Y) = \mathrm{Hom}_{\underline{B}}(B,SY)$$

such that (i) $FSY \to Y$ is an effective epimorphism for all $Y \in \mathrm{Ob}\underline{A}$, (ii) FB is projective for all $B \in \mathrm{Ob}\underline{B}$. These adjoint functors define a cotriple (see [3]) and hence cohomology groups $H^*_{cot}(X,A)$

defined by

$$H^*_{cot}(X,A) = H^*\{h_A(C.(X))\}$$

where $C.(X)$ is the simplicial object of $\underline{A}/X$ with $C_q(X) = (FS)^{q+1}(X)$ with face and degeneracy operators coming from the adjunction maps $id \to SF$, $FS \to id$.

(3) Suppose that $\underline{A}$ is closed under finite limits and has sufficiently many projective objects. Regarding X as a constant simplicial object there exists by Prop. 3, §4 a trivial fibration $P. \to X$, where $P.$ is cofibrant, which is unique up to homotopy over X . The group $H^q\{h_A(P.)\}$ is therefore independent of the choice of $P.$ and we denote it by $R^q h_A(X)$.

(4) Suppose that $\underline{A}$ satisfies the conditions of Theorem 4, §4, that the abelianization functor $ab:\underline{A}/X \to (\underline{A}/X)_{ab}$ exists, and that $(\underline{A}/X)_{ab}$ is an abelian category. Then the model category $\underline{C} = s(\underline{A}/X)$ satisfies the assumptions made at the beginning of this section and hence cohomology groups $H^*_M(X,A)$ are defined, where X and A are identified with constant simplicial objects.

<u>Theorem 5</u>: When each of the groups $H^q_M(X,A)$, $H^q_{cot}(X,A)$, and $R^q h_A(X)$ is defined, it is canonically isomorphic with the Grothendieck sheaf cohomology group $H^q_{GT}(X,A)$.

<u>Proof</u>: We begin by showing that $H^q_M(X,A) = R^q h_A(X)$. Let $\underline{F}$ be the abelian category $(A/X)_{ab}$. $\underline{F}$ has enough projectives, namely those of the form P_{ab} where P is a projective object

of $\underline{A}/X$. Hence $s\underline{F}$ and $Ch(\underline{F})$ are model categories (see Remark 5 at the end of §4) and $N: s\underline{F} \to Ch(\underline{F})$ is an equivalence of model categories. The loop and suspension functors on $Ho(Ch\underline{F})$ are given very simply by functors Ω and Σ on $Ch\underline{F}$ defined by the formulas

$$(\Sigma X)_q = \begin{cases} X_{q-1} & q > 0 \qquad d\Sigma x = -\Sigma dx \\ 0 & q = 0 \end{cases}$$

$$(\Omega X)_q = \begin{cases} X_{q+1} & q > 0 \qquad d\Omega x = -\Omega dx \\ \operatorname{Ker}\{d: X_1 \to X_0\} & q = 0 \end{cases}$$

Let $A[q]$ be the chain complex in $\underline{F}$ which is A in dimension q and 0 elsewhere ($A[q] = 0$ if $q < 0$) . As $NK(A,0) = A[0]$ $N\Omega^{q+N}\Sigma^N K(A,0) = A[q]$, hence $H^q_M(X,A) = [\underline{\underline{L}}\,ab(X), \Omega^{q+N}\Sigma^N K(A,0)] = \pi_0((P.)_{ab}, \Omega^{q+N}\Sigma^N K(A,0)) = \pi(N(P.)_{ab}, A[q]) = H^q \operatorname{Hom}_{\underline{F}}(N(P.)_{ab}, A) = H^q(\operatorname{Hom}_{\underline{F}}((P.)_{ab}, A)) = H^q h_A(P.) = R^q h_A(X)$.

To finish the theorem we need some results about Grothendieck sheaves ([1], [19]). Let $\underline{T}$ denote a Grothendieck topology whose underlying category is closed under finite projective limits and has sufficiently many projectives, and where a covering of an object Y in $\underline{T}$ is a family $\underline{U} = (Z \to Y)$ consisting of a single effective epimorphism. Eventually we will let $\underline{T}$ be $\underline{A}/X$. A <u>presheaf</u> of sets (resp. abelian groups) is a functor $\underline{T}^o \to (\text{sets})$ (resp. (ab)) and a sheaf of sets (resp. abelian groups) is a presheaf F such that for any effective epimorphism $Z \to Y$ the diagram

$$F(Y) \to F(Z) \rightrightarrows F(Z \times_Y Z)$$

is exact. Letting Pr and Sh (resp. Prab and Shab) denote the categories of presheaves and sheaves of sets (resp. abelian groups) we have the diagram

$$(4) \qquad \begin{array}{ccc} \underline{\mathrm{Sh}} & \overset{a}{\underset{i}{\leftrightarrows}} & \underline{\mathrm{Pr}} \\ \underline{Z}\downarrow\uparrow j & & \mathbb{Z}\downarrow\uparrow j \\ \underline{\mathrm{Shab}} & \overset{a}{\underset{i}{\leftrightarrows}} & \underline{\mathrm{Prab}} \end{array}$$

Here i and j are inclusion functors which are right adjoint functors and the other functors are left adjoint functors. The square of left (resp. right) adjoint functors commutes up to canonical isomorphisms.

We recall the construction of a , the associated sheaf functor. If $F \in \mathrm{Ob}\ \underline{\mathrm{Pr}}$ (resp. $\mathrm{Ob}\ \underline{\mathrm{Prab}}$), then the 0-th (resp. q-th) Cech cohomology presheaf of F is defined by

$$\underline{\check{H}}^0(F)(Y) = \varinjlim_{\underline{U}} \check{H}^0(\underline{U},F)$$

$$(\text{resp. } \underline{\check{H}}^q(F)(Y) = \varinjlim_{\underline{U}} \check{H}^q(\underline{U},F))$$

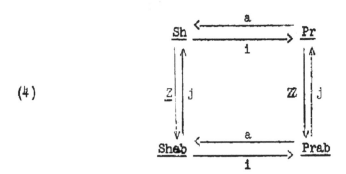

where the limit is taken over the category of coverings $\underline{U} = (U \to Y)$ of Y and where

$$\check{H}^o((U \to Y),F) = \operatorname{Ker}\{F(U) \rightrightarrows F(U \times_Y U)\}$$

(resp. $\check{H}^q((U \to Y),F)$ = the q-th cohomology of the cosimplicial abelian group

$$F(U) \rightrightarrows F(U \times_Y U) \text{⇶} F(U \times_Y U \times_Y U) \ \ldots \quad) .$$

Then $aF = \check{\underline{H}}^o\check{\underline{H}}^o(F)$. Given Y , choose an effective epimorphism $P \to Y$ with P projective; it follows that $(P \to Y)$ is cofinal in the category of coverings of Y and hence

$$\check{H}^o(F)(Y) = \operatorname{Ker}\{F(P) \rightrightarrows F(P \times_Y P)\}$$

In particular $\check{\underline{H}}^o(F)(P) = F(P)$ if P is projective, and hence

(5) $$(aF)(P) = F(P) .$$

If Y is arbitrary choose effective epimorphism $P_o \to Y$, $P_1 \to P_o \times_Y P_o$, whence

$$a(F)(Y) = \operatorname{Ker}\{(aF)(P_o) \rightrightarrows (aF)(P_1)\}$$

$$= \operatorname{Ker}\{F(P_o) \rightrightarrows F(P_1)\} .$$

It follows that for $F \in \operatorname{Ob} \underline{\text{Prab}}$, $aF = 0$ if and only if $F(P) = 0$ for all projective P . Now if $F' \xrightarrow{u} F \xrightarrow{v} F''$ are maps in $\underline{\text{Shab}}$

with $vu = 0$, then this sequence is exact iff $aH = 0$ where $H = \mathrm{Ker}\, v/\mathrm{Im}\, u$ in the category <u>Prab</u> . Hence we have proved

<u>Lemma 1</u>: A sequence $F' \to F \to F''$ of abelian sheaves is exact iff $F'(P) \to F(P) \to F''(P)$ is exact for all projective objects P .

Let $\mathbb{Z}(S)$ denote the free abelian group generated by a set S . Then the abelianization functor $\mathbb{Z}$ for presheaves is given by $(\mathbb{Z}F)(Y) = \mathbb{Z}(F(Y))$ for all Y hence combining (5) and the commutativity of (4) we obtain

<u>Lemma 2</u>: If F is a sheaf of sets, then its abelianization $\underline{\mathbb{Z}}F$ is such that

$$\underline{\mathbb{Z}}F(P) = \mathbb{Z}(F(P))$$

for all projectives P .

Let $\underline{H}^q$:<u>Shab</u> $\to$ <u>Prab</u> be the q-th cohomology presheaf functors. Then $\underline{H}^q$ $q \geq 0$ are the right derived functor of $i:\underline{S} \to \underline{P}$ and $\underline{H}^*(F)(Y) = H^*(Y,F)$ is the cohomology of F over Y . We define a weak equivalence in $s\underline{T}$ to be a map $Z. \to Y.$ such that for any projective object P, $\underline{\mathrm{Hom}}(P,Y.) \to \underline{\mathrm{Hom}}(P,Z.)$ is a weak equivalence in $\underline{S}$. This agrees with the definition in §4.

<u>Proposition 1</u>: The following are equivalent for a sheaf of abelian groups:

(i) $\underline{H}^q(F) = 0 \quad q > 0$.

(ii) $\check{H}^q((U \to Y),F) = 0 \quad q > 0$ for all effective epimorphisms

$U \to Y$.

(iii) For any weak equivalence $Z. \to Y.$ in $s\underline{T}$

$$H^*(F(Y.)) \cong H^*(F(Z.))$$

A sheaf satisfying the equivalent conditions of Prop. 1 will be called flask. By (i) any injective sheaf is flask.

Proof: (i) $\Longrightarrow$ (iii). Let $h_Y : \underline{T}^o \to (\text{sets})$ be the functor represented by Y; then h_Y is a sheaf. Let $\underline{Z}_Y = \underline{Z}h_Y$ so that

$$\text{Hom}_{\underline{\text{Shab}}}(\underline{Z}_Y, F) = F(Y) .$$

Let $I^{\cdot}$ be an injective resolution of F in Shab so that $H^q(I^{\cdot}(Y)) = H^q(Y,F) = 0$ for all Y. Then

$$H_h^p H_v^q \text{Hom}(\underline{Z}_{Y.}, I^{\cdot}) = H^p H^q(Y., F) = \begin{cases} H^p(F(Y.)) & q = 0 \\ 0 & q > 0 \end{cases}$$

$$H_v^p H_h^q \text{Hom}(\underline{Z}_{Y.}, I^{\cdot}) = H^p \text{Hom}(H_q(\underline{Z}_{Y.}), I^{\cdot}) = \text{Ext}^p(H_q(\underline{Z}_{Y.}), F)$$

and so we obtain a spectral sequence

$$E_2^{pq} = \text{Ext}^p(H_q(\underline{Z}_{Y.}), F) \Longrightarrow H^{p+q}(F(Y.))$$

and a similar spectral sequence for $Z.$. Hence we are reduced to showing that $H_*(\underline{Z}_{Z.}) \cong H_*(\underline{Z}_{Y.})$. By Lemmas 1 and 2 we are reduced to showing that $\mathbb{Z}\,\underline{\mathrm{Hom}}(P,Z.) \to \mathbb{Z}\,\underline{\mathrm{Hom}}(P,Y.)$ is a weak equivalence of simplicial abelian groups for each projective P . But this is clear since $\mathrm{Hom}(P,Z.) \to \underline{\mathrm{Hom}}(P,Y.)$ is a weak equivalence and since $\pi_*(\mathbb{Z}K.)$, the homology of a simplicial set $K.$, is a weak homotopy invariant.

(ii)$\Longrightarrow$ (i). There is a Cartan-Leray spectral sequence $E_2^{pq} = \check{\underline{H}}^p(\underline{H}^q F) \Longrightarrow \underline{H}^{p+q}F$ ([1] 3.5, Ch.I.). By assumption $E_2^{po} = \check{\underline{H}}^p F = 0$ for $p > 0$ hence by induction on n one sees that $\underline{H}^n F =$

(iii)$\Longrightarrow$ (ii). $\check{H}^q((U \to Y),F) = H^q(F(Z.))$ where $Z.$ is the object of $s\underline{T}$ with $Z_q = U \times_Y \cdots \times_Y U$ $q+1$ times. Regarding Y as a constant simplicial object, $Z. \to Y$ is a weak equivalence. In effect if P is projective $\mathrm{Hom}(P,U) \to \mathrm{Hom}(P,Y)$ is surjective; denoting this by $S \to T$ we have that $\underline{\mathrm{Hom}}(P,Z.) \to \underline{\mathrm{Hom}}(P,Y.)$ is the map

$$\begin{array}{ccccc} \cdots\ S\times_T S\times_T S & \rightrightarrows & S\times_T S & \rightrightarrows & S \\ \downarrow & & \downarrow & & \downarrow \\ \cdots\ T & \rightrightarrows & T & \rightrightarrows & T \end{array}$$

which is a homotopy equivalence by the cone construction. Q.E.D.

Lemma 3: With the notations of (2) $C.(X) \to X$ is a weak equivalence.

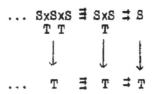

Proof: Let P be projective, as $FSP \to P$ is an effective epimorphism it follows that P is a retract of FSF. It suffices to show therefore that $\underline{\mathrm{Hom}}_A(FB, C.(X)) \to \underline{\mathrm{Hom}}_A(FB, X)$ or $\underline{\mathrm{Hom}}_B(B, SC.(X)) \to \underline{\mathrm{Hom}}_B(B, SX)$ is a weak equivalence of simplicial sets. However $SC.(X) \to SX$ is a homotopy equivalence by the "cone construction". Q.E.D.

We can now finish the proof of the theorem. Let $\underline{T} = \underline{A}/X$ and let $I^{\cdot}$ be a flask resolution of the sheaf h_A and let $P. \to X$ be a weak equivalence where each P_q is projective. For the double complex $I^{\cdot}(P.)$ we have

$$H_h^p H_v^q(I^{\cdot}(P.)) = \begin{cases} H^p(I^{\cdot}(X)) & q = 0 \\ 0 & q > 0 \end{cases} = \begin{cases} H^p_{GT}(X,A) & \\ 0 & q > 0 \end{cases}$$

by Prop. 1 and

$$H_v^p H_h^q(I^{\cdot}(P.)) = \begin{cases} H^p(h_A(P.)) & q = 0 \\ 0 & q > 0 \end{cases} = \begin{cases} R^p h_A(X) & q = 0 \\ 0 & q > 0 \end{cases}$$

by Lemma 1. Thus the two spectral sequences of a double complex degenerate giving $H^p_{GT}(X,A) \simeq R^p h_A(X)$. Similarly $H^p_{GT}(X,A) \simeq H^p_{cot}(X,A)$ by Lemma 3 and condition (ii) on the functors (3). Q.E.D.

Example: Let $\underline{A}$ = (gps.) and let G be a group. Then any abelian group object in $\underline{A}/G$ is of the form $M \times_\tau G \xrightarrow{pr_2} G$ where

M is a G module and $M \times_T G$ denotes the semi-direct product of M and G. Hence $(\underline{A}/G)_{ab}$ is the abelian category of G modules. Moreover if $X \to G$ is a group over G, then $\mathrm{Hom}_{\underline{A}/G}(X, M\times_T G) = \mathrm{Der}(X,M)$, the derivations of X with values in M regarded as an X module via the map $X \to G$. For each group X over G, let $C^q(X,M) = \mathrm{Hom}_{sets}(X^q,M)$ be the group of q cochain of X with values in M and let $\delta : C^q(X,M) \to C^{q+1}(X,M)$ be the usual coboundary operator. Then

$$0 \to \mathrm{Der}(\cdot,M) \to C'(\cdot,M) \xrightarrow{\delta} C^2(\cdot,M) \xrightarrow{\delta} \cdots$$

is a flask resolution of the sheaf $h_{M\times_T G}$ on $\underline{A}/G$. In effect any weak equivalence of simplicial groups is a homotopy equivalence of sets and the functors $C^q(X,M)$ depend only on the underlying set of X; hence $C^q(\cdot,M)$ is flask by Prop. 1(iii). On the other hand the sequence is exact by Lemma 1 and the fact that the cohomology of a free group vanishes in dimension ≥ 2. Thus we find that

$$H^q_{GT}(G,M) = H^q_{cot}(G,M) = H^q_{M}(G,M) = \begin{cases} H^{q+1}(G,M) & q \geq 1 \\ \mathrm{Der}(G,M) & q = 0 \end{cases}$$

where $H^*(G,M)$ is ordinary group cohomology.

<u>Remarks</u>: 1. The preceding example generalizes immediately to cover the usual cohomology of Lie algebras and associative algebras over a field ([3]). Moreover one is lead to the following general

picture for cohomology of any kind of universal algebras. Letting $\underline{A}$ be a category of universal algebras and $X \in \mathrm{Ob}\underline{A}$, then an X module is an abelian group object A in $\underline{A}/X$, and the cohomology of X with values in A may be defined to be either $H^*_M(X,A)$, $H^*_{GT}(X,A)$, or $H^*_{cot}(X,A)$ where the cotriple is for example the "underlying set" and "free algebra" functors $\underline{A} \leftrightarrows$ (Sets). A cochain complex for computing this cohomology is just a flask resolution of the sheaf h_A on $\underline{A}/X$.

2. The isomorphism $H^*_{GT}(X,A) = H^*(h_A(P.))$ is a special case of a general theorem of Verdier that the Grothendieck sheaf cohomology group $H^q(X,F)$ may be computed as $\varinjlim_{\underline{U}} H^*(\underline{U},F)$ where $\underline{U}$ runs over the category of hypercoverings of X for the topology. In effect $P. \to X$ is cofinal in this category of hypercoverings. See [18] especially, exposé V, appendice.

§6. Simplicial modules over a simplicial ring.

In this section we show how the category $\underline{M}_R$ of left simplicial modules over a simplicial ring forms closed simplicial model category. $\underline{M}_R$ occurs as the category $(s\underline{A}/X)_{ab}$ where X is a non-constant simplicial object in $s\underline{A}$ and hence is worth studying in virtue of §5. We also derive Künneth spectral sequences which are useful in applications. Some applications to simplicial groups are given.

In this section a ring is always associative with unit, not necessarily commutative, and left or right modules are always unitary.

Let R be a simplicial ring. By a left simplicial R module we mean a simplicial abelian group M together with a map $R \times M \to M$ of simplicial sets which for each q makes M_q into a left R_q module. The left simplicial R modules form an abelian category $\underline{M}_R$ where a sequence is exact iff it is exact in each dimension. The category of right simplicial R modules is the category $\underline{M}_{R^o}$ where R^o is the simplicial ring which is the simplicial abelian group R with the multiplication opposed to that of R.

If $X,Y \in \mathrm{Ob}\,\underline{M}_R$, let $\underline{\mathrm{Hom}}_R(X,Y)_n = \mathrm{Hom}_{\underline{M}_R}(X \otimes_{\mathbb{Z}} \mathbb{Z}\Delta(n), Y)$ with simplicial operator φ^* induced by $\tilde{\varphi}$ in the obvious way. Here $\mathbb{Z}K$ denotes the simplicial abelian group obtained by applying the free abelian group functor dimension-wise to the simplicial set K and $\otimes$ denotes dimension-wise tensor product. There is a bilinear map

(1) $$\underline{\mathrm{Hom}}_R(X,Y)\otimes\underline{\mathrm{Hom}}_R(Y,Z)\to\underline{\mathrm{Hom}}_R(X,Z)$$

defined by letting $g\circ f$ for $f: X\otimes\mathbb{Z}\Delta(n)\to Y$ and $g: Y\otimes\mathbb{Z}\Delta(n)\to Z$ be the map

$$X\otimes\mathbb{Z}\,\Delta(n)\xrightarrow{id\otimes\Delta}X\otimes\mathbb{Z}\,\Delta(n)\otimes\mathbb{Z}\,\Delta(n)\xrightarrow{f\otimes id}Y\otimes\mathbb{Z}\,\Delta(n)\xrightarrow{g}Z\ .$$

It is clear that $\underline{M}_R$ is a simplicial category with $\underline{\mathrm{Hom}}_{\underline{M}_R}(X,Y)$ = the underlying simplicial set of $\underline{\mathrm{Hom}}_R(X,Y)$ and with composition induced by (1). If K is a simplicial set, let $X\otimes_{\mathbb{Z}}\mathbb{Z}K$ and $\underline{\mathrm{Hom}}_{\underline{S}}(K,Y)$ be considered as simplicial R modules in the natural way. Then there are canonical isomorphisms

$$\mathrm{Hom}_{\underline{S}}(K,\underline{\mathrm{Hom}}_R(X,Y)) = \mathrm{Hom}_{\underline{M}_R}(X\otimes_{\mathbb{Z}}\mathbb{Z}K,Y) = \mathrm{Hom}_{\underline{M}_R}(X,\underline{\mathrm{Hom}}_{\underline{S}}(K,X))$$

(2) $$X\otimes_{\mathbb{Z}}\mathbb{Z}(K\times L) = (X\otimes_{\mathbb{Z}}\mathbb{Z}K)\otimes_{\mathbb{Z}}\mathbb{Z}L$$

$$\underline{\mathrm{Hom}}_{\underline{S}}(K\times L,Y) = \underline{\mathrm{Hom}}_{\underline{S}}(L,\underline{\mathrm{Hom}}_{\underline{S}}(K,Y))$$

which may be used as in the proof of Prop. 2, §1/to show that $X\otimes_{\mathbb{Z}}\mathbb{Z}K$ is an object $X\otimes K$ and that $\underline{\mathrm{Hom}}_{\underline{S}}(K,Y)$ is an object Y^K in the simplicial category $\underline{M}_R$. We will use the notation $X\otimes\mathbb{Z}K$ instead of $X\otimes K$ in the following.

Define a map in $\underline{M}_R$ to be a fibration (resp. weak equivalence) if it is so as a map in $\underline{S}$, and call a map a cofibration if it has the LLP with respect to the trivial fibrations. The proof that $\underline{M}_R$ is a closed simplicial model category follows that for $\underline{G}$

(§3) and $s\underline{A}$ in the case (*) (§4); in effect every object is fibrant and factorization axiom may be proved by the small object argument. The following descriptions hold: A map $f:X \to Y$ in $\underline{M}_R$ is a fibration if $(f,\epsilon):X \to Y \times_{K(\pi_o Y,0)} K(\pi_o X,0)$ is surjective, a weak equivalence if $\pi_* f:\pi_* X \xrightarrow{\sim} \pi_* Y$, and a trivial fibration if f is a surjective weak equivalence. f is a cofibration iff it is a retract of a free map, and a trivial cofibration iff f is a cofibration and a strong deformation retract map. Here $f:X \to Y$ is said to be <u>free</u> if there are subsets $C_q \subset Y_q$ for each q such that C_* is stable under the degeneracy operators of Y and $X_q \oplus R_q C_q \xrightarrow{\sim} Y_q$ for each q.

If A is a ring, $\underline{M}_A$ is the category of left A modules, and R is the constant simplicial ring obtained from A, then $\underline{M}_R = s(\underline{M}_A)$ and the above structure of a closed simplicial model category on $\underline{M}_R$ is the same as that defined in §4. Moreover if $Ch(\underline{M}_A)$ denotes the category of chain complexes in $\underline{M}_A$, then the normalization functor $N:\underline{M}_R \to Ch(\underline{M}_A)$ is an equivalence of closed model categories. Here $Ch(\underline{M}_A)$ is defined to be a closed model category by a slight modification of example B, §1, Ch.I. The following fact is of course clear for $Ch(\underline{M}_A)$.

<u>Proposition 1</u>: Let Ω and Σ be the loop and suspension functors in the category $Ho(\underline{M}_R)$. Then

$$\theta: M \xrightarrow{\sim} \Omega\Sigma M$$

$$\Sigma\Omega M \xrightarrow{\sim} M \Longleftrightarrow \pi_o M = 0$$

where the maps are adjunction morphisms. Furthermore if $A' \xrightarrow{i} A \xrightarrow{j} A'' \xrightarrow{\delta} \Sigma A'$ is a cofibration sequence in $Ho\underline{M}_R$, then $\Omega\Sigma A' \xrightarrow{-i\theta^{-1}} A \xrightarrow{j} A'' \xrightarrow{\delta} \Sigma A'$ is a fibration sequence.

<u>Proof</u>: For any simplicial left R module X there are canonical exact sequences in $\underline{M}_R$

(3) $$0 \to X \to CX \to \Sigma X \to 0$$

(4) $$0 \to \Omega X \to \Lambda X \to X \to K(\pi_0 X,0) \to 0$$

which in more detail are the maps

$$X \xrightarrow{i_1} X\otimes\mathbb{Z}\,\Delta(1)/X\otimes\mathbb{Z}\,\{0\} \longrightarrow X\otimes\mathbb{Z}\Delta(1)/X\otimes\mathbb{Z}\dot{\Delta}(1)$$

$$0\times_X X^{\Delta(1)}\times_X 0 \xrightarrow{(pr_1,pr_2)} 0\times_X X^{\Delta(1)} \xrightarrow{j_1 pr_2} X \xrightarrow{\epsilon} K(\pi_0 X,0)$$

Here $K(\pi_0 X,0)$ is the simplicial R module which is the constant simplicial abelian group of $\pi_0 X$ with R module structure determined via $\epsilon:R \to K(\pi_0 R,0)$ and the natural $\pi_0 R$ action on $\pi_0 X$, and $\epsilon:X \to K(\pi_0 X,0)$ is the canonical augmentation. The exactness of (3) is clear dimension-wise and (4) is exact for simplicial groups hence also for $\underline{M}_R$, since $X^{\Delta(1)}$ is calculated in $\underline{M}_R$ as in $\underline{G}$. The canonical homotopy $h:\Delta(1)\times\Delta(1) \to \Delta(1)$ with $hi_0 = i_0\sigma$ and $hi_1 = id$ induces a homotopy $H:CX\otimes\mathbb{Z}\Delta(1) \to CX$ with $Hi_0 = 0$ $Hi_1 = id$ and a homotopy $K:\Lambda X\otimes\mathbb{Z}\Delta(1) \to \Lambda X$ with $Ki_0 = 0$ and $Ki_1 = id$. Hence $\pi(CX) = \pi(\Lambda X) = 0$.

The functor Ω on $\underline{M}_R$ defined by (4) actually becomes the functor Ω in $Ho(\underline{M}_R)$, since every X in $\underline{M}_R$ is fibrant and so $X^{\Delta(1)}$ is a path object for X . Similarly one sees that ΣX represents the suspension of X in $Ho(\underline{M}_R)$ provided X is cofibrant. However if $Y \to X$ is a trivial fibration with Y cofibrant we obtain a map into (3) of the corresponding sequence for Y , so by the homotopy long exact sequence and 5 lemma $\Sigma Y \to \Sigma X$ is a weak equivalence. Therefore ΣX represents the suspension of X in $Ho(\underline{M}_R)$ for all X .

If $\pi_0 X = 0$, then the diagram

$$(5) \qquad \begin{array}{ccccccccc} 0 & \longrightarrow & \Omega X & \longrightarrow & C\Omega X & \longrightarrow & \Sigma\Omega X & \longrightarrow & 0 \\ & & \downarrow{\scriptstyle 1} & & \downarrow{\scriptstyle u} & & \downarrow{\scriptstyle v} & & \\ 0 & \longrightarrow & \Omega X & \longrightarrow & \Lambda X & \longrightarrow & X & \longrightarrow & 0 \end{array} ,$$

where u and v are induced by the contracting homotopy K of ΛX described above, and the five lemma show that v is a weak equivalence. However v is the adjunction map for the adjoint functors Σ and Ω in $\underline{M}_R$ and hence also in $Ho(\underline{M}_R)$, so the direction $\Longleftarrow$ of the second assertion of the proposition is proved. The direction $\Longrightarrow$ results from the formula $\pi_0(\Sigma X) = 0$ which follows since $(\Sigma X)_0 = 0$. The first assertion of the proposition may be proved by a diagram similar to (5). For the last assertion of the proposition we may assume that $i : A' \to A$ is a cofibration of cofibrant objects, that A'' is the cone on i , that j is the embedding of

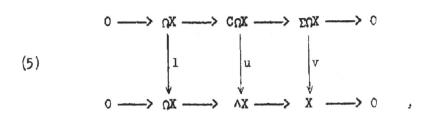

A as the base of this cone, and finally that δ is the cokernel of j. As $\underline{M}_R$ is abelian δ is a fibration with fiber $j: A \to A''$ and there is a diagram

$$\begin{array}{ccccccc} A' & \xrightarrow{i} & A & \xrightarrow{j} & A'' & \xrightarrow{\delta} & \Sigma A' \\ \downarrow{\scriptstyle -\theta} & & \downarrow{\scriptstyle 1} & & \downarrow{\scriptstyle 1} & & \downarrow{\scriptstyle 1} \\ \Omega\Sigma A' & \overset{\partial}{\dashrightarrow} & A & \xrightarrow{j} & A'' & \xrightarrow{\delta} & \Sigma A' \end{array}$$

where ∂ is the boundary operator of the fibration sequence associated to δ. For the commutativity of the first square see proof of Prop. 6, §3, Ch.I. As θ is an isomorphism we find that $\partial = -i\theta^{-1}$ and so the proposition is proved.

<u>Kunneth spectral sequences</u>. If X and Y are simplicial abelain groups and if $x \in X_p$, $y \in Y_q$, then the element $x \underline{\otimes} y \in (X \otimes Y)_{p+q}$ is defined by the formula

$$x \underline{\otimes} y = \sum_{(\mu,\nu)} \varepsilon(\mu,\nu) s_\nu x \otimes s_\mu y \tag{6}$$

where (μ,ν) runs over all p,q shuffles, i.e. permutations $(\mu_1 \ldots \mu_p, \nu_1, \ldots, \nu_q)$ of $[0,\ldots,p+q-1]$ such that $\mu_1 < \ldots < \mu_p$ and $\nu_1 < \ldots < \nu_q$, where $\varepsilon(\mu,\nu)$ is the sign of the permutation, and where $s_\mu y = s_{\mu_p} \ldots s_{\mu_1} y$, $s_\nu x = s_{\nu_q} \ldots s_{\nu_1} x$. The following properties of the operation $\underline{\otimes}$ are well known.

(1) $x \in NX, y \in NY \Longrightarrow x\otimes y \in N(X\otimes Y)$

(2) $d(x\underline{\otimes}y) = dx\underline{\otimes}y + (-1)^p x\underline{\otimes}dy$ where p = degree x and $d = \Sigma(-1)^i d_i$.

(3) $x\underline{\otimes}(y\underline{\otimes}z) = (x\underline{\otimes}y)\underline{\otimes}z$

(4) If $\tau: X\otimes Y \cong Y\otimes X$ is the isomorphism $\tau(x\otimes y) = y\otimes x$, then $\tau(x\underline{\otimes}y) = (-1)^{pq}\tau(y\underline{\otimes}x)$ if p = degree x, q = degree y.

If R is a simplicial ring, then these properties show that $\underline{\otimes}$ induces on NR the structure of a differential graded ring which is anti-commutative if R is commutative. In fact NR is even strictly anti-commutative ($x^2 = 0$ if degree x is odd) when R is commutative as one sees directly from (6). Consequently $\pi_*R = H_*(NR)$ is a graded ring which is strictly anti-commutative if R is commutative. If X is a left (resp. right) simplicial R module then by virtue of $\otimes$, NX is a left (resp. right) differential graded NR module, and so π_*X is a left (resp. right) graded π_*R module.

By a <u>projective resolution</u> of a left simplicial R module X we mean a trivial fibration $u:P \to X$ in $\underline{M}_R$ such that P is cofibrant. By Prop. 4, §2 u is unique up to homotopy over X , and moreover if we choose projective resolutions $p_Y:Q(Y) \to Y$ for each $Y \in Ob\underline{M}_R$ and a map $Q(f)$ for each map $f:Y \to Y'$ such that $p_{Y'}Q(f) = fp_Y$, then we obtain a functor $\pi_o(\underline{M}_R) \to \pi_o(\underline{M}_{R,c})$ right adjoint to the inclusion functor. Hence projective resolution is up to homotopy a homotopy preserving functor of X .

If X is a right simplicial R module and Y is a left simplicial R module, and if $P \xrightarrow{u} X$ and $Q \xrightarrow{v} Y$ are projective resolutions of X and Y in $\underline{M}_{R^o}$ and $\underline{M}_R$ respectively, then the abelian group $P\otimes_R Q$ is independent up to homotopy over $X\otimes_R Y$ of the choices of u and v. We denote $P\otimes_R Q$ by $X\overset{L}{\otimes}_R Y$ and call it the <u>derived tensor product</u> of X and Y since in the terminology of §4, Ch.I it is the total left derived functor of $\otimes_R : \underline{M}_{R^o} \times \underline{M}_R \to \underline{M}_{\mathbb{Z}}$.

<u>Theorem 6</u>: Let R be a simplicial ring and let X (resp. Y) be a left (resp. right) simplicial R module. Then there are canonical first quadrant spectral sequences

(a) $E^2_{pq} = \pi_p(\mathrm{Tor}^R_q(X,Y)) \Longrightarrow \pi_{p+q}(X\overset{L}{\otimes}_R Y)$

(b) $E^2_{pq} = \mathrm{Tor}^{\pi R}_p(\pi M,\pi N)_q \Longrightarrow$ "

(c) $E^2_{pq} = \pi_p(\pi_q X\overset{L}{\otimes}_R Y) \Longrightarrow$ "

(d) $E^2_{pq} = \pi_p(X\overset{L}{\otimes}_R \pi_q Y) \Longrightarrow$ "

which are functorial in R,X,Y.

In (a) $\mathrm{Tor}^R_q(X,Y)$ denotes the simplicial abelian group obtained by applying the functor $\mathrm{Tor}^{\cdot}_q(\cdot,\cdot)$ to R,X,Y dimensionwise. In (b) $\mathrm{Tor}^{\pi R}_p(\pi M,\pi N)_q$ denotes the homogeneous submodule of degree q in $\mathrm{Tor}^{\pi R}_p(\pi M,\pi N)$ which is naturally graded since the ring πR and the modules $\pi M,\pi N$ are graded. In (c) $\pi_q X$ is an abbreviation for the constant simplicial abelian group $K(\pi_q X,0)$ which becomes a right R module via the augmentation $R \to K(\pi_o R,0)$ and the action $\pi_q X\otimes\pi_o R \to \pi_q X$ induced by $\underline{\otimes}$. Similarly for $\pi_q Y$ in (d).

Proof: (a) Construct recursively an exact sequence in $\underline{M}_R$

$$\ldots \to P_1 \to P_0 \to X \to 0 \tag{7}$$

by letting $X_0 = X$, $P_q \to X_q$ be a projective resolution of X_q, and $X_{q+1} = \mathrm{Ker}(P_q \to X_q)$. Then $\pi P_q = 0$ for $q > 0$ so $P_q \to 0$ is a weak equivalence of cofibrant objects and hence a homotopy equivalence. Hence there is a map $h: P_q \otimes_{\mathbb{Z}} \mathbb{Z}\Delta(1) \to P_q$ with $h(\mathrm{id}\otimes i_0) = \mathrm{id}$ and $h(\mathrm{id}\otimes i_1) = 0$. Thus $(P_q \otimes_R Y) \otimes_{\mathbb{Z}} \mathbb{Z}\Delta(1) \otimes_R Y \xrightarrow{h\otimes \mathrm{id}} P_q \otimes_R Y$ is a contracting homotopy of $P_q \otimes_R Y$ and so $\pi(P_q \otimes_R Y) = 0$. Think of the simplicial operators in (7) as being vertical and consider the double complex $N_*^v(P_* \otimes_R Y)$ obtained by applying the normalization functor to the simplicial structure. Then $H_p^h H_q^v = 0$ for $q > 0$ and $= \pi_p(P_0 \otimes_R Y)$ if $q = 0$. As the cofibrant simplicial R module P_q is a direct summand of a free simplicial R module, $(P_q)_n$ is projective over R_n for each n, and so in simplicial dimension n (7) is a projective resolution of the R_n module X_n. Hence $H_q^h N_n^v(P_* \otimes_R Y) = N_n^v H_q^h(P_* \otimes_R Y) = N_n \mathrm{Tor}_q^{R_n}(X_n, Y_n)$, where we have used that N is an exact functor from simplicial abelian groups to chain complexes.

Thus we obtain the spectral sequence

$$E_{pq}^2 = \pi_p(\mathrm{Tor}_q^R(X,Y)) \Longrightarrow \pi_{p+q}(P_0 \otimes_R Y) \tag{8}$$

having the edge homomorphism $\pi_n(P_0 \otimes_R Y) \to \pi_n(X \otimes_R Y)$ induced by the map $P_0 \to X$. By repeating this procedure with Y instead of X we obtain a spectral sequence

$$(9) \qquad E^2_{pq} = \pi_p(\mathrm{Tor}^R_q(X,Y)) \Longrightarrow \pi_{p+q}(X\otimes_R Q_o)$$

where $Q_o \to Y$ is a projective resolution of Y, whose edge homomorphism $\pi_n(X\otimes_R Q_o) \to \pi_n(X\otimes_R Y)$ is induced by v. Substituting P_o for X in (9), it degenerates showing that $P_o\otimes_R Q_o \to P_o\otimes_R Y$ is a weak equivalence and hence that $\pi(X\overset{L}{\otimes}_R Y) = \pi(P_o\otimes_R Y)$. Substituting this into (8) we obtain the spectral sequence (a) and the following fact which will be used later.

<u>Corollary</u>: The edge homomorphism $\pi(X\overset{L}{\otimes}_R Y) \to \pi(X\otimes_R Y)$ of spectral sequence (a) is induced by the canonical map $X\overset{L}{\otimes}_R Y \to X\otimes_R Y$. This map is a weak equivalence if $\mathrm{Tor}^{R_n}_q(X_n,Y_n) = 0$ for $q > 0$, $n \geq 0$.

To prove (a) is functorial let $R,X,Y \to R',X',Y'$ be a map and suppose that a sequence (7)' corresponding to (7) has been constructed. As a map of simplicial modules is a trivial fibration iff it is so as a map of simplicial sets, the maps $P'_q \to X'_q$ are trivial fibrations as maps in $\underline{M}_{R^o}$. Hence we may construct a map θ from (7) to (7)' covering the given map $X \to X'$ by inductively defining $\theta_q : P_q \to P'_q$ by lifting in

(10)

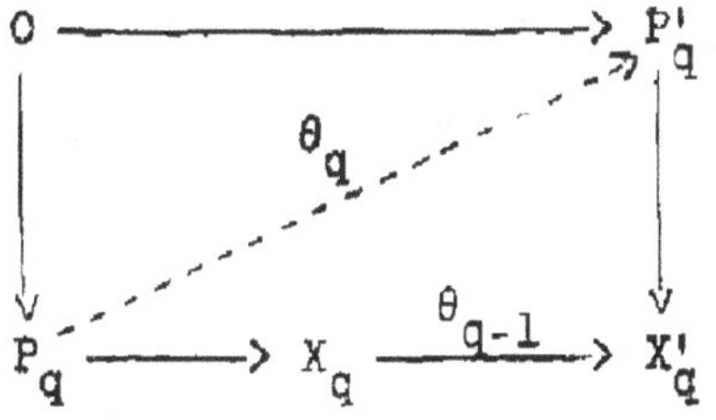

We then obtain a map of the spectral sequence (8) into the corresponding one (8)' which is independent of the choice of θ because its E^2 term is clearly independent and the map $P_o \to P'_o$ covering $X \to X'$ is unique up to homotopy. Consequently there is a canonical map from spectral sequence (a) to the corresponding one (a)' and this proves the functorality of (a) as well as its independence of the choices made for its construction.

(b) We need two lemmas.

Lemma 1: Suppose that P is a cofibrant right simplicial R module such that $\pi_* P$ is a free graded $\pi_* R$ module. Then for any left simplicial R module Y the map

$$\pi_* P \underset{\pi_* R}{\otimes} \pi_* Y \to \pi_* (P \otimes_R Y)$$

induced by $\underline{\otimes}$ is an isomorphism.

Lemma 2: Suppose P is as in Lemma 1 and let $f : X \to Y$ be a fibration in $\underline{M}_{R^o}$ such that $\pi_* f : \pi_* X \to \pi_* Y$ is surjective. Then given any map $u : P \to Y$ there is a $v : P \to X$ with $fv = u$.

Proof: Let $RS^n = \operatorname{Coker}(R \otimes \mathbb{Z}\,\dot{\Delta}(n) \to R \otimes \mathbb{Z}\,\Delta(n))$ $n \geq 1$ considered as a right simplicial R module in the obvious way, let $t_n \in (RS^n)_n$ be the residue class of the element $1 \otimes \mathrm{id}_{[n]}$, and let u_n be the element of $\pi_n(RS^n)$ represented by t_n. We claim that

A. $\pi(RS^n)$ is a free right graded πR module generated by u_n

B. The map $\pi(RS^n)\otimes_{\pi R}\pi Y \to \pi(RS^n\otimes_R Y)$ induced by $\underset{\sim}{\otimes}$ is an isomorphism.

Indeed there is an exact sequence of right simplicial R modules

$$(11) \qquad 0 \to RS^{n-1} \overset{i}{\to} RD^n \overset{j}{\to} RS^n \to 0$$

where $RD^n = \mathrm{Coker}(R\otimes\mathbb{Z}\,V(n,0) \to R\otimes\mathbb{Z}\,\Delta(n))$, where i is induced by $\tilde{\partial}_0 : \Delta(n-1) \to \Delta(n)$ and j is the canonical surjection. Moreover $0 \to RD^n$ is a trivial cofibration because it is a cobase extension of the map $R\otimes\mathbb{Z}\,V(n,0) \to R\otimes\mathbb{Z}\,\Delta(n)$, which is a trivial cofibration by SM7 since R is cofibrant. Hence RD^n is contractible and the long exact sequence in homotopy yields an isomorphism

$$\pi_q(RS^n) \xrightarrow[\sim]{\partial} \begin{cases} \pi_{q-1}(RS^{n-1}) & q \geq 1 \\ 0 & q = 0 \end{cases}$$

such that $\partial u_n = u_{n-1}$. By property (2) of $\underset{\sim}{\otimes}$ ∂ is an isomorphism $\pi(RS^n) \cong \Sigma\pi(RS^{n-1})$ of right graded πR modules, where if M is a right graded module over a graded ring S, we define ΣM to be the right graded R module with $(\Sigma M)_k = M_{k-1}$ and $(\Sigma m)s = \Sigma(ms)$; here if $m \in M_{k-1}$, Σm denotes m as an delement of $(\Sigma M)_k$. A then follows by induction on n. To obtain B note that (11) splits in each dimension so it remains exact after tensoring with Y over R. The resulting long exact homotopy sequence yields the bottom isomorphism in the square

$$\begin{array}{ccc}
\pi(RS^n)\otimes_{\pi R}\pi Y & \xrightarrow{\partial\otimes id} & \Sigma\pi(RS^{n-1})\otimes_{\pi R}\pi Y \\
\downarrow & & \downarrow \\
\pi(RS^n\otimes_R Y) & \xrightarrow[\sim]{\partial} & \Sigma\pi(RS^{n-1}\otimes_R Y)
\end{array}$$

where the vertical arrows come from $\underline{\otimes}$ and the diagram commutes by property (2) of $\underline{\otimes}$. Induction on n then proves B.

If P is as in Lemma 1 choose elements $x_i \in P_{n_i}$, $i \in I$ whose representatives in πP form a free basis over πR, and let $\varphi: \oplus RS^{n_i} \to P$ the map of right simplicial R modules sending t_{n_i} to x_i. By the assumption on P and A φ is a weak equivalence hence a homotopy equivalence since both are cofibrant. Lemma 1 then reduces to the case $P = RS^n$ in which case it follows from B.

To prove Lemma 2 we reduce by the covering homotopy theorem to the case $P = RS^n$, and we must show that $Z_n f: Z_n X \to Z_n Y$ is surjective where Z_n denotes the group of n cycles in the normalization. As f is a fibration $N_j f$ is surjective $j > 0$ and as πf is surjective one sees easily that Zf is surjective. Q.E.D.

To obtain (b) construct an exact sequence

(12) $$\to P_1 \to P_0 \to X \to 0$$

of right simplicial R modules by setting $X_0 = X$, $X_{q+1} = \mathrm{Ker}(u_q: P_q \to X_q)$ where u_q is surjective, πu_q is surjective, and

πP_q is a free graded right πR module. u_q may be obtained by choosing generators $\{\alpha_i\}$ for πX_q over πR, letting $v:\oplus_i RS^{n_i} \to X_q$ be a map sending t_{n_i} onto a representative for α_i, and then factoring $v = u_q i$ where u_q is a fibration and i is a trivial fibration. If $Q \to Y$ is a projective resolution of Y, consider the double complex $N_*^v(P_* \otimes_R Q)$ where v refers to the (vertical) simplicial structure. By Lemma 1 $\pi(P_q \otimes_R Q) = \pi P_q \otimes_{\pi R} \pi Q$ and by the construction of (12), $\pi(P_*)$ is a free πR resolution of πX. Thus $H_p^h H_q^v(N_*(P_* \otimes_R Q)) = H_p^h(\pi P_* \otimes_{\pi R} \pi Q)_q = \mathrm{Tor}_p^{\pi R}(\pi X, \pi Q)_q$. On the other hand, Q is projective over R in each dimension, hence $H_p^v H_q^h(N_*(P_* \otimes_R Q)) = H_p^v N_* H_q^h(P_* \otimes_R Q) = 0$ if $q > 0$ and $\pi_p(X \otimes_R Q)$ if $q = 0$. As $\pi(X \otimes_R Q) \simeq \pi(X \overset{L}{\otimes}_R Y)$ by the above corollary and $\pi(Q) = \pi(Y)$ we obtain spectral sequence (b) from the two spectral sequences of a double complex. The functorability of (b) as well as its independence of (12) may be proved in exactly the same way as for (a), except the lifting analogous to (10) is constructed via Lemma 2.

(c) These are derived by the Serre-Postnikov method. In effect we have (see Prop. 1 (4)) canonical exact sequences

$$0 \to \Omega X \to \Lambda X \to X \to \pi_0 X \to 0 \tag{13}$$

in $\underline{M}_{R^0}$, where ΛX is contractible and where $\pi_0 X$ is short for the right simplicial R module which is the constant simplicial abelian group $K(\pi_0 X, 0)$, and whose R module structure comes via the augmentation $\varepsilon:R \to K(\pi_0 R,0)$ from the map $\pi_0 X\otimes\pi_0 R \to \pi_0 X$ induced by Φ. From the long exact homotopy sequence we have

(14) $$\pi_q(X) \xrightarrow[\sim]{\partial} \pi_{q-1}(\Omega X) \qquad q > 0$$

where $\partial(\alpha\cdot\rho) = (\partial\alpha)_\rho$ if $\rho \in \pi R$. Hence substituting $\Omega^k X$ into we obtain exact sequences

(15) $$0 \to \Omega^{k+1}X \to \Lambda\Omega^k X \to \Omega^k X \to \pi_k X \to 0 \qquad k \geq 0$$

where $\pi_k X$ stands for the right simplicial R module as described in the theorem. Letting $Q \to Y$ be a projective resolution of Y, $\otimes_R Q$ is exact and $\Lambda\Omega^k X\otimes_R Q$ is contractible, hence from (15) we obtain exact sequences

$$\to \pi_{n-1}(\Omega^{k+1}X\otimes_R Q) \to \pi_n(\Omega^k X\otimes_R Q) \to \pi_n(\pi_k X\otimes_R Q) \to \pi_{n-2}(\Omega^{k+1}X\otimes_R Q) \to \cdots$$

for $k \geq 0$. By the corollary $\otimes_R Q$ may be replaced by $\overset{L}{\otimes}_R Y$ and so we obtain an exact couple (D^2_{pq}, E^2_{pq}) with $E^2_{pq} = \pi_p(\pi_q X\overset{L}{\otimes}_R Y)$ and $D^2_{pq} = \pi_p(\Omega^q X\overset{L}{\otimes}_R Y)$ and hence the spectral sequence (c). It is clearly canonical and functorial since the only choice made was that of Q which is unique and functorial up to homotopy. Spectral sequence (d) is proved similarly. There is no sign trouble from the fact that $\partial:\pi_q Y \to \pi_{q-1}(\Omega Y)$ satisfies $\partial(\rho\alpha) = (-1)^k\rho\cdot\partial\alpha$ if $\rho \in \pi_k R$ because only $k = 0$ occurs when we consider $\pi_k Y$ as a left simplicial R module. Theorem 1 is now proved. Q.E.D.

Applications to simplicial groups. Let G be a simplicial group. If M is a simplicial G module we call $H_*(G,M) = \pi(\mathbb{Z} \overset{L}{\otimes}_{\mathbb{Z}G} M)$ the homology of G with coefficients in M. Here $\mathbb{Z}$ is short for $K(\mathbb{Z},0)$ with trivial G action. To calculate the homology we choose a projective resolutive of $\mathbb{Z}$ as a right $\mathbb{Z}G$ module, e.g. $\mathbb{Z}WG$ where $WG \to \overline{W}G$ is the universal principal G bundle, whence $H(G,M) = \pi(\mathbb{Z}WG \otimes_{\mathbb{Z}G} M)$. If M is an abelian group on which $\pi_0 G$ acts and we consider M as a constant simplicial G module, then it follows that $H_*(G,M)$ is the homology of the simplicial set $\overline{W}G$ with values in the local coefficient system defined by M. In particular when G is a constant simplicial group and M is a G module $H(G,M)$ in the above sence coincides with the ordinary group homology of G with values in M.

If F is a free group, then

$$\mathrm{Tor}_q^{\mathbb{Z}F}(\mathbb{Z},\mathbb{Z}) = \begin{cases} \mathbb{Z} & q = 0 \\ F_{ab} & q = 1 \\ 0 & q \geq 2 \end{cases}$$

hence if G is a simplicial group which is free in each dimension spectral sequence (a) degenerates giving

(16) $$H_n(G,\mathbb{Z}) = \begin{cases} \mathbb{Z} & n = 0 \\ \pi_{n-1}(G_{ab}) & n > 0 \end{cases}$$

which is a formula due to Kan [10] when G is a free simplicial group.

Let $f:G \to H$ be a weak equivalence of simplicial groups. Then f is a weak equivalence in $\underline{S}_f$ and as every object of $\underline{S}$ is cofibrant f is a homotopy equivalence in $\underline{S}$. Thus $\mathbb{Z}G \to \mathbb{Z}H$ is a homotopy equivalence of simplicial abelian groups and so $\pi\mathbb{Z}G \cong \pi\mathbb{Z}H$. From spectral sequence (b) we deduce that $H_*(G,\mathbb{Z}) \cong H_*(H,\mathbb{Z})$ which shows that homology is a weak homotopy invariant.

Suppose now that $1 \to K \to G \to H \to 1$ is an exact sequence of simplicial groups and that M is a simplicial G module. Let $P \to M$ be a projective resolution of M as on left $\mathbb{Z}G$ module. Then

$$\mathbb{Z} \otimes_{\mathbb{Z}G} M \simeq \mathbb{Z} \otimes_{\mathbb{Z}G} P \simeq \mathbb{Z} \otimes_{\mathbb{Z}H} (\mathbb{Z}H \otimes_{\mathbb{Z}G} P)$$

and

$$\pi_q(\mathbb{Z}H \otimes_{\mathbb{Z}G} P) \simeq \pi_q(\mathbb{Z} \otimes_{\mathbb{Z}K} P) = H_q(K,M) \quad .$$

Substituting $R = \mathbb{Z}H$, $X = \mathbb{Z}$, $Y = \mathbb{Z} \otimes_{\mathbb{Z}K} P$ in spectral sequence (d) we obtain a spectral sequence

$$(17) \qquad E^2_{pq} = H_p(H, H_q(K,M)) \Longrightarrow H_{p+q}(G,M)$$

which generalizes the Hochshild-Serre spectral sequence for group homology and the Serre spectral sequence for the fibration $\overline{W}K \to \overline{W}G \to \overline{W}H$.

Spectral sequence (a) with $R = \mathbb{Z}G$, $X = \mathbb{Z}$, $Y = \mathbb{Z}$ has the edge homomorphism

$$H_n(G,\mathbb{Z}) \to \pi_{n-1}(G_{ab}) \qquad n > 0$$

which is an isomorphism for $n = 1$ in general and for all n if G is free. So we obtain Poincare's theorem

$$H_1(G, \mathbb{Z}) = (\pi_0 G)_{ab} \quad .$$

Now by the method of Serre [16] it is possible to start from this fact and the spectral sequence (17) and prove directly the Hurewicz and Whitehead theorems for simplicial groups. We leave the details to the reader.

Bibliography

[1] Artin, M.: Grothendieck topologies (mimeographed notes). Harvard, 1962.

[2] Artin, M. and B. Mazur: Homotopy of Varieties in the Etale Topology (to appear).

[3] Barr, M. and J. Beck: Acyclic models and triples, La Jolla Conf. on Categorical Algebra (to appear).

[4] Barratt, M.G., V.K.A.M. Guggenheim, and J.C. Moore: On semi-simplicial fibre fundles. Am. J. Math. 81, 639-657 (1959).

[5] Dold, A.: Homology of Symmetric Products and Other Functors of Complexes. Ann. of Math. 68, 54-80 (1958).

[6] Dold, A., and D. Puppe: Homologie nicht-additver Funktoren; Anwendungen. Ann. Inst. Fourier 11, 201-312 (1961).

[7] Gabriel, P. and Zisman, M.: Calculus of Fractions and Homotopy Theory, Springer, Berlin (1966)

[8] Gresheshen, H.H.: Higher composition products, J. Math. Kyoto Univ. 5, No. 1, 1-37 (1965).

[9] Hartshorne, R.: Residues and Duality. Lecture Notes in Mathematics No. 20, Springer, Berlin (1966).

[10] Kan, D.M.: On homotopy theory and C.S.S. groups. Ann. of Math. 68, 38-53 (1958).

[11] ________ The Hurewicz theorem. Proc. Int. Symp. Algebraic Topology and its applications, Mexico 1956.

[12] ________ On c.s.s. complexes. Am. J. Math. 79, 449-476 (1957).

[13] ________ On c.s.s. categories. Bol. Soc. Math. Mexicana 1957, 82-94.

[14] Lawvere, F.W.: Functorial semantics of algebraic theories, Proc. Nat. Acad. Sci. U.S.A., 50, 869-872 (1963).

[15] Milnor, J.: The geometric realization of a semi-simplicial complex. Ann. of Math. 65, 357-362 (1957).

[16] Serre, J.-P.. Groupes d'homotopy et classes de groupes abélians. Ann. of Math. 58, 258-294 (1953).

[17] Spanier, E.: Higher Order Operations. Trans. Am. Math. Soc., 109, 509-539 (1963).

[18] Verdier, J.-L.: Séminaire de Géométrie algébrique, (1963-64)--Cohomology etale des Schémas. Exposes I-V (Mimeographed notes, Institute des Hautes Études Scientifiques).

[19] Verdier, J.-L.: Categories derivées.- mimeographed notes, Institute des Hautes Études Scientifiques.

Lecture Notes in Mathematics

Bisher erschienen/Already published

Vol. 1: J. Wermer, Seminar über Funktionen-Algebren. IV, 30 Seiten. 1964. DM 3,80

Vol. 2: A. Borel, Cohomologie des espaces localement compacts d'après J. Leray. IV, 93 pages. 1964. DM 9,–

Vol. 3: J. F. Adams, Stable Homotopy Theory. 2nd. revised edition. IV, 78 pages. 1966. DM 7,80

Vol. 4: M. Arkowitz and C. R. Curjel, Groups of Homotopy Classes. 2nd. revised edition. IV, 36 pages. 1967. DM 4,80

Vol. 5: J.-P. Serre, Cohomologie Galoisienne. Troisième édition. VIII, 214 pages. 1965. DM 18,–

Vol. 6: H. Hermes, Eine Termlogik mit Auswahloperator. IV, 42 Seiten. 1965. DM 5,80

Vol. 7: Ph. Tondeur, Introduction to Lie Groups and Transformation Groups. VIII, 176 pages. 1965. DM 13,50

Vol. 8: G. Fichera, Linear Elliptic Differential Systems and Eigenvalue Problems. IV, 176 pages. 1965. DM 13.50

Vol. 9: P. L. Ivănescu, Pseudo-Boolean Programming and Applications. IV, 50 pages. 1965. DM 4,80

Vol. 10: H. Lüneburg, Die Suzukigruppen und ihre Geometrien. VI, 111 Seiten. 1965. DM 8,–

Vol. 11: J.-P. Serre, Algèbre Locale. Multiplicités. Rédigé par P. Gabriel. Seconde édition. VIII, 192 pages. 1965. DM 12,–

Vol. 12: A. Dold, Halbexakte Homotopiefunktoren. II, 157 Seiten. 1966. DM 12,–

Vol. 13: E. Thomas, Seminar on Fiber Spaces. IV, 45 pages. 1966. DM 4,80

Vol. 14: H. Werner, Vorlesung über Approximationstheorie. IV, 184 Seiten und 12 Seiten Anhang. 1966. DM 14,–

Vol. 15: F. Oort, Commutative Group Schemes. VI, 133 pages. 1966. DM 9,80

Vol. 16: J. Pfanzagl and W. Pierlo, Compact Systems of Sets. IV, 48 pages. 1966. DM 5,80

Vol. 17: C. Müller, Spherical Harmonics. IV, 46 pages. 1966. DM 5,–

Vol. 18: H.-B. Brinkmann und D. Puppe, Kategorien und Funktoren. XII, 107 Seiten. 1966. DM 8,–

Vol. 19: G. Stolzenberg, Volumes, Limits and Extensions of Analytic Varieties. IV, 45 pages. 1966. DM 5,40

Vol. 20: R. Hartshorne, Residues and Duality. VIII, 423 pages. 1966. DM 20,–

Vol. 21: Seminar on Complex Multiplication. By A. Borel, S. Chowla, C. S. Herz, K. Iwasawa, J.-P. Serre. IV, 102 pages. 1966. DM 8,–

Vol. 22: H. Bauer, Harmonische Räume und ihre Potentialtheorie. IV, 175 Seiten. 1966. DM 14,–

Vol. 23: P. L. Ivănescu and S. Rudeanu, Pseudo-Boolean Methods for Bivalent Programming. 120 pages. 1966. DM 10,–

Vol. 24: J. Lambek, Completions of Categories. IV, 69 pages. 1966. DM 6,80

Vol. 25: R. Narasimhan, Introduction to the Theory of Analytic Spaces. IV, 143 pages. 1966. DM 10,–

Vol. 26: P.-A. Meyer, Processus de Markov. IV, 190 pages. 1967. DM 15,–

Vol. 27: H. P. Künzi und S.T. Tan, Lineare Optimierung großer Systeme. VI, 121 Seiten. 1966. DM 12,–

Vol. 28: P. E. Conner and E. E. Floyd, The Relation of Cobordism to K-Theories. VIII, 112 pages. 1966. DM 9.80

Vol. 29: K. Chandrasekharan, Einführung in die Analytische Zahlentheorie. VI, 199 Seiten. 1966. DM 16.80

Vol. 30: A. Frölicher and W. Bucher, Calculus in Vector Spaces without Norm. X, 146 pages. 1966. DM 12,–

Vol. 31: Symposium on Probability Methods in Analysis. Chairman: D.A. Kappos. IV, 329 pages. 1967. DM 20,–

Vol. 32: M. André, Méthode Simpliciale en Algèbre Homologique et Algèbre Commutative. IV, 122 pages. 1967. DM 12,–

Vol. 33: G. I. Targonski, Seminar on Functional Operators and Equations. IV, 112 pages. 1967. DM 10,–

Vol. 34: G. E. Bredon, Equivariant Cohomology Theories. VI, 64 pages. 1967. DM 6,80

Vol. 35: N. P. Bhatia and G. P. Szegö, Dynamical Systems: Stability Theory and Applications. VI, 416 pages. 1967. DM 24,–

Vol. 36: A. Borel, Topics in the Homology Theory of Fibre Bundles. VI, 95 pages. 1967. DM 9,–

Bitte wenden/Continued

Vol. 37: R. B. Jensen, Modelle der Mengenlehre. X, 176 Seiten. 1967. DM 14,–

Vol. 38: R. Berger, R. Kiehl, E. Kunz und H.-J. Nastold, Differentialrechnung in der analytischen Geometrie. IV, 134 Seiten. 1967. DM 12,–

Vol. 39: Séminaire de Probabilités I. IV, 187 pages. 1967. DM 14,–

Vol. 40: J. Tits, Tabellen zu den einfachen Lie Gruppen und ihren Darstellungen. VI, 53 Seiten. 1967. DM 6,80

Vol. 41: R. Hartshorne, Local Cohomology. VI, 106 Seiten. 1967. DM 10,–

Vol. 42: J. F. Berglund and K. H. Hofmann, Compact Semitopological Semigroups and Weakly Almost Periodic Functions. VI, 146 Seiten. 1967. DM 12,–